CAMBRIDGE COUNTY GEOGRAPHIES

SCOTLAND

General Editor: W. Murison, M.A.

ABERDEENSHIRE

Cambridge County Geographies

ABERDEENSHIRE

by

ALEXANDER MACKIE, M.A.

Late Examiner in English, Aberdeen University, and
author of *Nature Knowledge in Modern Poetry*

With Maps, Diagrams and Illustrations

Cambridge:
at the University Press

1911

CAMBRIDGE UNIVERSITY PRESS
Cambridge, New York, Melbourne, Madrid, Cape Town,
Singapore, São Paulo, Delhi, Mexico City

Cambridge University Press
The Edinburgh Building, Cambridge CB2 8RU, UK

Published in the United States of America by Cambridge University Press, New York

www.cambridge.org
Information on this title: www.cambridge.org/9781107653528

First published 1911
First paperback edition 2013

A catalogue record for this publication is available from the British Library

ISBN 978-1-107-65352-8 Paperback

CONTENTS

ILLUSTRATIONS

The illustrations on pp. 3, 12, 62, 63 are from photographs
by W. Norrie; those on pp. 5, 9, 14, 19, 21, 22, 23, 24, 26, 28,
30, 31, 32, 34, 36, 48, 53, 56, 58, 60, 61, 67, 86, 93, 97, 98, 99,

100, 101, 103, 120, 127, 128, 129, 130, 133, 140, 141, 144, 146, 147, 149, 150, 151, 152, 153, 155, 157, 158, 159, 161, 182, 183, 184, 186, 188 and 189, are from photographs by J. Valentine and Sons; that on p. 7 from a photograph by J. Watt; that on p. 47 from a photograph by Dr W. Brown; that on p. 84 from a photograph by A. Gordon; that on p. 193 from a photograph by A. Gray.

Thanks are due to W. Duthie, Esq., Collynie, for permission to reproduce the illustration on p. 82; to J. M^cG. Petrie, Esq., Glen-Logie, for permission to reproduce that on p. 81; to Messrs T. and R. Annan and Sons, for permission to reproduce that on p. 175; to the Society of Antiquaries of Scotland for permission to reproduce those on pp. 113 and 118; and to Alexander Walker, Jr., Esq., Aberdeen, for permission to reproduce that on p. 179.

1. County and Shire. The Origin of Aberdeenshire.

The term "shire," which means a division (Anglo-Saxon *sciran*: to cut or divide), has in Scotland practically the same meaning as "county." In most cases the two names are interchangeable. Yet we do not say Orkney-shire nor Kirkcudbrightshire. Kirkcudbright is a stewartry and not a county, but in regard to the others we call them with equal readiness shires or counties. County means originally the district ruled by a Count, the Norman equivalent of Earl. It is said that Aberdeenshire is the result of a combination of two counties, Buchan and Mar, representing the territory under the rule of the Earl of Buchan and the Earl of Mar. The distinction is in effect what we mean to-day by East Aberdeenshire and West Aberdeenshire ; and the local students of Aberdeen University when voting for their Lord Rector by "nations" are still classified as belonging to either the Buchan nation or the Mar nation according to their place of birth.

The counties, then, are certain areas which it is convenient for political and administrative purposes to

divide the country into for the better and more convenient
management of local and internal affairs. To-day
Scotland has thirty-three of these divisions. In a public
ordinance dated 1305, twenty-five counties are named.
They would seem to have been first defined early in the
twelfth century, but as a matter of fact nothing very
definite is known, either as to the date of their origin or
as to the principles which regulated the making of their
geographical boundaries. It is certain, however, that the
county divisions were in Scotland an introduction from
England. The term came along with the people who
were flocking into Scotland from the south. The lines
were drawn for what seemed political convenience and
no doubt they were suited to the times. To-day the
boundaries seem on occasion somewhat erratic. Banchory,
for example, is in Kincardineshire, while Aboyne and
Ballater on the same river bank and on the same line of
road and railway are in Aberdeenshire. If the carving
were to be done over again in the twentieth century,
more consideration would probably be given to the railway
lines.

A commission of 1891 did actually rearrange the
boundaries. Of the parishes partly in Aberdeen and
partly in Banff, some were transferred wholly to Aberdeen (Gartly, Glass, New Machar, Old Deer and
St Fergus), while others were placed in Banffshire
(Cabrach, Gamrie, Inverkeithny, Alvah and Rothiemay).
How it happened that certain parts of adjoining counties
were planted like islands in the heart of Aberdeenshire
may be understood by reference to such a case as that of

St Fergus. A large part of this parish belonged to the Cheynes, who being hereditary sheriffs of Banffshire were naturally desirous of having their patrimonial estates under their own legal jurisdiction, and were influential enough to be able to stereotype this anomaly. This explains the place of St Fergus in Banffshire; it is now very properly a part of Aberdeenshire.

The lone Kirkyard, Gamrie

The county took its name from the chief town—Aberdeen—which is clearly Celtic in origin and means the town at the mouth of either the Dee or the Don. Both interpretations are possible; but the fact that the Latin form of the word has always been *Aberdonia* and *Aberdonensis*, favours the Don as the naming river. As a matter of fact, Old Aberdeen, though lying at no great distance from the bank of the Don, can hardly be said to be associated with Donmouth, whereas a considerable population must from a remote period have been located at the mouth of the Dee. Whatever interpretation is accepted, it was this city—the only town in the district conspicuous for population and resources—that gave its name to the county as a whole.

The whole region between the river Dee and the river Spey, comprising the two counties of Banff and Aberdeen, forms a natural province. There is no natural, or recognisable line of demarcation between the two counties. Their fortunes have been one. The river Deveron might conceivably have been chosen as the dividing line, but in practice it is so only to a limited extent. The whole district, which if invaded was never really conquered by the Romans, made one of the seven Provinces of what was called Pictland in the early middle ages, and it long continued to assert for itself a semi-independent political existence.

2. General Characteristics.

The county is almost purely agricultural. It has always enjoyed a certain measure of maritime activity and of recent years the fishing industry, especially at Aberdeen, has made immense progress, but as a whole the area is a well-cultivated district. Round the coast and on all the lower levels tillage is the rule. In the interior the level of the land rises rapidly, and ploughed fields

Town House, Old Aberdeen

give place to desolate moors and bare mountain heights in which agriculture is an impossible industry. The surface of the lowland parts, now in regular cultivation, was originally very rough and rock-strewn. It was covered with erratic blocks of stone, gneiss and granite (locally called "heathens"), left by the melting of the ice fields which overspread all the north-east of Scotland during the Ice Age. These stones have been cleared from the fields and utilised as boundary walls. Some idea of the extra-ordinary energy and excessive labour necessary to clear the land for tillage may be gathered from a glance at the "consumption" dyke at Kingswells, some five miles from Aberdeen. This solid rampart stretches like a great break-water across nearly half a mile of country, through a dip to the south of the Brimmond Hill. It is five or six feet in height and twenty to thirty in breadth and contains thousands of tons of troublesome boulders gathered from the surrounding slopes. The disposal of these blocks was a serious problem. It has been solved by this rampart. In other parts the stones were built up into enclosing walls and now serve the double purpose of enclosing the fields and providing a certain amount of shelter for crops and cattle. The slopes of the Brimmond Hill are in certain parts still uncleared and the appearance of these areas helps us to realise what this section of the country looked like before the enterprising agriculturist braced himself to prepare the surface for the use of the plough.

The soil, except in the alluvial deposits on the banks of the Don and the Ythan, is not of great natural fertility, yet by the exceptional industry of the inhabitants and their enterprise as a farming community it has been raised to a high degree of productiveness. The county now enjoys a well but hard earned reputation for progressive agriculture. Notably so in regard to cattle-breeding. It

Consumption Dyke at Kingswells

is the home of a breed of cattle called Aberdeenshire, black and polled, but it is just as famous for its strain of shorthorns which have been bred with skill and insight for more than a century. In spite, then, of its inferior soil, its wayward climate and its northern latitude, the inborn stubbornnesss and determination of its people have made it a great and prosperous agricultural region and only those who on a September day have seen from the top of Benachie the undulating plains of Buchan glittering golden in the sun can realise what a transformation has been effected on a barren and stony land by the industry of man.

The most easterly of the Scottish counties, it abuts like a prominent shoulder into the North Sea. It has, therefore, a considerable sea-board partly flat and sandy, partly rocky and precipitous. The population of the numerous villages dotted along this coast used in time past to devote themselves to fishing, but the tendency of recent years has been to concentrate this industry in the larger towns, Fraserburgh, Peterhead and Aberdeen.

Other industries there are few. Next to agriculture and fishing comes granite, which is the only mineral worthy of mention found in the county. It is the prevailing rock of the district and is quarried to a considerable extent in various parts. A large part of the population earn their living by this industry, and Aberdeen granite, like Aberdeen beef and Aberdeen fish, is a well-known product and travels far. Paper and wool are also manufactured but only on a moderate scale.

There is only one other general feature of the county that deserves mention and that is its attractiveness as a health resort. The banks of the Dee, more especially in its upper regions, is a much frequented holiday haunt; and every summer and autumn Braemar, Ballater and Aboyne are crowded with visitors from all parts of the country. The late Queen Victoria no doubt gave the

The Punch Bowl, Linn of Quoich, Braemar

impetus to this fashion. Her majesty at an early period
of her reign bought the estate of Balmoral, half-way
between Ballater and Braemar, and having built a royal
castle there made it her practice to reside for a large part
of every year amongst the Deeside hills. Apart from this
royal advertisement the high altitude of the district, and
its dry, bracing climate, as well as its romantic mountain
scenery, have proved permanently attractive. Here are
Loch-na-gar (sung by Byron), Ben-Macdhui, Brae-riach,
Ben-na-Buird, Ben-Avon and other Bens, all of them
4000, or nearly 4000, feet above sea-level, and all of
them imposing and impressive in their bold and massive
forms. These mountains supply elements of grandeur
which exercise a fascination upon people who habitually
live in a flat country, and Braemar is not likely to lose its
merited popularity.

3. Size. Shape. Boundaries.

Aberdeenshire is one of the large counties in area,
standing fifth in Scotland. Although Inverness contains
more than twice the number of square miles in Aberdeenshire, its population is far behind that of Aberdeen, which
in this respect is the third county in Scotland. Its
greatest length from N.E. to S.W. is 102 miles; its
greatest breadth from N.W. to S.E. is 50. The coast
line measures 65 miles and is little indented. The whole
area of the county is 1970 square miles, or 1,261,971
acres, of which 6400 are water.

In shape the county might be likened to a pear lying obliquely on its side, the narrow stalk-end being in the mountains, while the rounded bulging head is the north-eastern sea-board. The flattest portion is the region lying north of the Ythan, called Buchan, and even this can hardly be called flat, for the level is broken by Mormond Hill, near Strichen, rising to a height of 810 feet. All the way to Pennan Head the contour of the land is irregularly wavy. The narrower portion in the S.W., called Mar, is entirely mountainous, and midway between these two extremes lie the Garioch and Formartin— districts which are undulating in character. A crescent line drawn from Aberdeen to Turriff, the convex side being to the S.W., would divide the county into two parts, which might be described as lowland and highland. The lowland portion contains the lower valley of the Don as far up as Inverurie, the valley of the Ythan and all the remaining northern part of the county. South of this imaginary line the ground rises in ridge after ridge until it culminates in the lofty Grampian range of the Cairn-gorms. The bipartite character of the county, which is reflected in the occupation and pursuits as in the character and language of the two populations, is of some import-ance, and yet must not be pressed too far, because the population in the one half is practically insignificant as compared with that of the other. It follows that when Aberdeenshire men and Aberdeenshire ways are referred to, nine times out of ten it is the lowland part of the county that is in question.

The boundaries are, on the east, the North Sea, and

Pennan, looking N.W.   Showing old and new houses of Troup

on the north as far west as Pennan Head, the Moray
Firth. There Banffshire and Aberdeenshire meet. From
that point inland a wavy boundary separates the two
counties, the Deveron being for part of the way the
dividing line. Above Rothiemay the boundary mounts
the watershed between Deveronside and Speyside, and
keeping irregularly to this line past the Buck of the
Cabrach, and the upper waters of the Don, reaches Ben-
Avon. Thence the line moves on to Ben-Macdhui with
Loch Avon on the right, and at Brae-riach Banffshire
ceases to be the boundary. For several miles, almost due
south in direction, Inverness comes in as the county on
the west. The southern boundary touches three counties,
Perth, Forfar and Kincardine. At Cairn Ealar, which is
the angle of turning and almost a right angle, the direction changes and runs east alongside of Perthshire to
the Cairnwell Road, and crossing this leaves Perthshire
at Glas Maol, where it touches Forfarshire. The line
continues east but with a trend to the north, passing on
the left Glenmuick, Glentanar and the Forest of Birse,
in which the Feugh takes its rise. On the top of Mount
Battock three counties meet, Forfar, Aberdeen and
Kincardine. Henceforth we are alongside of Kincardineshire and the line bends north-west with a semi-circular
sweep round Banchory-Ternan and the Hill of Fare to
Crathes, from a little beyond which, the bed of the Dee
becomes the boundary line all the way to Aberdeen. In
all this area of high ground the line of march is practically
the watershed throughout, marking off the drainage area
of the Don and the Dee from that of the Deveron and

Loch Avon and Ben-Macdhui

the Avon (a tributary of the Spey) on the one hand, and from that of the Tay and the two Esks (south and north) on the other.

4.   Surface, Soil and General Features.

From what has been already said of the contour of the county it may be inferred that its surface is extremely varied.   Every variety of highland and lowland country is to be found within its limits.   Near the sea-board the land is gently undulating, never quite flat but not rising to any great height till Benachie (1440) is reached. From that point onwards, whether up Deveronside or Donside or Deeside, the mountains rise higher and higher till the Cairngorms, which comprise some of the loftiest mountains in the kingdom, are reached.   At that point we are more than half-way across Scotland, and in reality are nearer to the Atlantic than to the North Sea.   Less than half the land is under cultivation.   Woods and plantations occupy barely a sixth part of the uncultivated area.   The rest is mountain and moor, yielding a scanty pasturage for sheep and red-deer, and on the lower elevations for cattle.

In the fringe round the sea-board no trees will grow. It is only when several miles removed from exposure to the fierce blasts that come from the North Sea that they begin to thrive, but the whole Buchan district is conspicuously treeless.   Almost every acre is cultivated and the succession of fields covered with oats, turnips and grass, which fill up the landscape as with a great patch-work, is broken only here and there by belts of trees round some manor-house or farm-steading. Except in a few places the scenery of this lowland portion is devoid of picturesque interest, yet the woods of Pitfour and of Strichen, the policies of Haddo House near Methlick, the quiet silvan beauty of Fyvie, which more resembles an English than a Scotch village, the wooded ridge that overlooks the Ythan at the Castle of Gight, are charming spots that serve by contrast to accentuate the general tameness of this lower area.

In the higher region, the south-western portions of the county, agriculture is, to some extent, practised, but it is necessarily confined to narrow strips in the valleys of the rivers. The hills, which are rarely wooded, and that only up to fifteen hundred feet above sea-level, are rounded in shape, not sharp and jagged. They are, where composed of granite, invariably clothed in heather and are occasionally utilised for the grazing of sheep, but this is becoming less common, and year by year larger areas are depleted of sheep for the better protection of grouse. All the heathery hills up to 2000 feet are grouse moors. Throughout the summer these display the characteristic brown tint of the heather—a tint which gives place in early August to a rich purple when the heather breaks into flower. Long strips of the heather-mantle are systematically burned to the ground every spring. Such blackened patches scoring with their irregular outlines the sides of the hills in April and May give a certain amount of variety to the prevailing tint of brown. They serve a very useful purpose. The young grouse shelter in the long and unburnt heather but frequent the cleared areas for the purpose of feeding on the tender young shoots which spring up from the blackened roots of the burned plants.

Further inland still, where the hills rise to a greater height, they become deer-forests. As a rule these forests are without trees and are often rockstrewn, bare and grassless. It is only in the sheltered corries or by the sides of some sparkling burn, that natural grasses spring up in sufficient breadth to provide summer pasturage for the red-deer, which are carefully protected for sporting purposes. Here too the ptarmigan breed in considerable numbers. The grouse moors command higher rents than would be profitable for a sheep-farmer to give for the grazing, and every year prior to the 12th of August, when grouse-shooting begins, there is an influx of sportsmen from the south, to enjoy this particular form of sport. The red grouse is indigenous to Scotland; it seems to find its natural habitat amongst the heather, where in spite of occasional failures in the nesting season, and in spite of many weeks' incessant shooting, it thrives and multiplies. Deer-stalking begins somewhat later; in a warm and favourable summer, the stags are in condition early in September. This sport is confined to a comparative few.

The highest mountain in the Braemar district is Ben-Macdhui (4296 feet). A few others are over 4000—Brae-riach and Cairntoul. Ben-na-Buird and Ben-Avon, which last is notable for the numerous tors or warty knots along its sky-line, are just under 4000 feet. Loch-na-gar, a few miles to the east and a conspicuous background to Balmoral Castle, is 3789. Byron called it "the most sublime and picturesque of the Caledonian Alps," and Queen Victoria writing from Balmoral in 1850 described it as "the jewel of all the mountains here." Its contour lines, which are somewhat more sharply curved than is usual in the Deeside hills, and the well-balanced distribution of its great mass make it easily recognised from a wide distance. This partly explains the pre-eminence which notwithstanding its inferiority of height it undoubtedly possesses. Due north from Ballater are Morven (2880) and Culblean, and due south is Mount Keen; a little east and on the boundary line of three counties is Mount Battock. Perhaps the most prominent hill, and the one most frequently visible to the great majority of Aberdeenshire folks, is Benachie, which stands as a fitting outpost of the vast regiment of hills. It stands apart and although only 1440 feet in height is an unfailing landmark from all parts of Buchan, from Aberdeen, from Donside, and even from Deveronside. Its well-defined outline and projecting "mither tap" render it an object of interest from far and near, while the presence or absence of cloud on its head and shoulders serves as a barometric index to the state of the weather.

Benachie

5. Watershed. Rivers. Lochs.

As we have already pointed out, the watershed co-incides to a large extent with the boundary line of the county. The lean of Aberdeenshire is from west to east so that all the rivers flow in an easterly direction to the North Sea. On the west and north-west of the highest mountain ridges, the slope of the land is to the north-east, and the Spey with its several tributaries carries the rainfall to the heart of the Moray Firth.

The chief river of the county is the Dee. It is the longest, the fullest-bodied, the most picturesque of all Aberdeenshire waters. Taking its rise in two small streams which drain the slopes of Brae-riach, it grows in volume and breadth, till, after an easterly course of nearly 100 miles, it reaches the sea at Aberdeen. The head-stream is the Garrachorry burn, which flows through the cleft between Brae-riach and Cairntoul. A more romantic spot for the cradle of a mighty river could hardly be found. The mountain masses rise steep, grim and imposing—on one side Cairntoul conical in shape, on the other Brae-riach broad and massive, a picture of solidity and immobility. The Dee well is 4060 feet above sea-level and 1300 above the stream which drains the eastern side of the Larig—the high pass to Strathspey. As it emerges from the Larig, it is a mere mountain torrent but presently it is joined at right angles by the Geldie from the south-west, and the united waters move eastward through a wild glen of rough and rugged slopes and ragged, gnarled Scots firs to the Linn of Dee, $6\frac{3}{4}$ miles above Braemar.   There is no great fall at the Linn, but here the channel of the river becomes suddenly contracted by great masses of rock and the water rushes through a

Linn of Dee, Braemar

narrow gorge only four feet wide.   The pool below is deep and black and much overhung with rocks.   For 300 yards stretches this natural sluice, formed by rocks with rugged sides and jagged bottom, the water racing past in small cascades.   The river is here spanned by a handsome granite bridge opened in 1857 by Queen Victoria.

As the river descends to Braemar, the glen gradually widens out, and the open, gravelly, and sinuous character of the bed, which is a feature from this point onwards, is very marked.   Pool and stream, stream and pool succeed

Old bridge of Dee, Invercauld

one another in shingly bends, clean, sparkling and beautiful.   At Braemar the bed is 1066 feet above sea-level. Below Invercauld the river is crossed by the picturesque old bridge built by General Wade, when he made his well-known roads through the Highlands after the rebellion of 1745.   Here the Garrawalt, a rough and obstructed

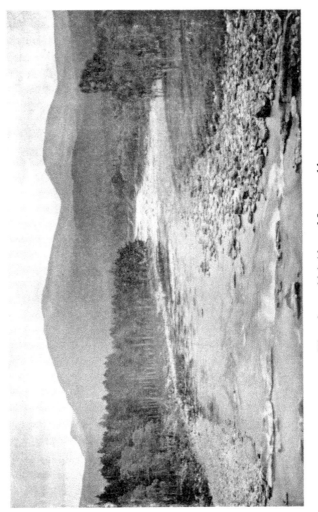

View from old bridge of Invercauld

tributary, joins the main river. From Invercauld past Balmoral Castle to Ballater is sixteen miles. Here the bottom is at times rocky, at times filled with big rough stones, at other times shingly but never deep. The average depth is only four feet, and the normal pace under ordinary conditions 3½ miles an hour. From Ballater,

Falls of Muick, Ballater

where the river is joined by the Gairn and the Muick, the Dee maintains the same character to Aboyne and Banchory, where it is joined by the Feugh from the forest of Birse. Just above Banchory is Cairnton, where the water supply for the town of Aberdeen, amounting on an average to 7 or 8 million gallons a day, is taken off. The course of the river near the mouth was diverted

some 40 years ago to the south, at great expense, by the
Town Council, and in this way a considerable area of
land was reclaimed for feuing purposes.   The spanning
of the river at this point by the Victoria bridge, which
superseded a ferry-boat, has led to the rise of a moderately
sized town (Torry) on the south or Kincardine side of the
river.

The scenery of Deeside, all the way from the Cairngorms to the old Bridge of Dee, two miles west of the
centre of the city, is varied and attractive.   It is well-
wooded throughout; in the upper parts the birch, which
would seem to be indigenous in the district, adds to the
beauty of the hill-sides, while the clean pebbly bed of the
river and its swift, dashing flow delight the eyes of those
who are familiar only with sluggish and mud-stained
waters.   It is not surprising therefore that the district
has attained the vogue it now enjoys.

The Don runs parallel to the Dee for a great part
of its course, but it is a much shorter river, measuring
only 78 miles.   It rises at the very edge of the county
close to the point where the Avon emerges from Glen
Avon and turns north to join the Spey.   It drains a valley
which is only ten or fifteen miles separated from the
valley of the larger river.   In its upper reaches it somewhat resembles Deeside, being quite highland in character;
but lower down the river loses its rapidity, becoming sluggish and winding.   Strathdon, as the upper area is called,
is undoubtedly picturesque, but it lacks the bolder features
of Deeside, being less wooded and graced with few hills
on the grand scale.   It has not, therefore, become a popular

Birch Tree at Braemar

summer resort, but its banks form the richest alluvial
agricultural land in the county—

> A mile o' Don's worth twa o' Dee
> Except for salmon, stone and tree.

This old couplet is so far correct.   The Dee is a great
salmon river, providing more first-class salmon angling
than any other river of Scotland, while the Don, though
owing to its muddy bottom a stream excellent beyond
measure and unsurpassed for brown trout, is not now,
partly owing to obstruction and pollution, a great salmon
river.   But the agricultural land on Donside, which for
the most part is rich deep loam, about Kintore, Inverurie
and the vale of Alford is much more kindly to the farmer
than the light gravelly soil of Deeside, which is so apt to
be burnt up in a droughty summer.   In the matter of
stone, things have changed since the couplet took shape.
The granite quarries of Donside are now superior to any
on the Dee; but the trees of Deeside still hold their own,
the Scots firs of Ballochbuie forest, west of Balmoral,
being the finest specimens of their kind in the north.

   The nether-Don has been utilised for more than a
century as a driving power for paper and wool mills.
Of these there is a regular succession for several miles
of the river's course, from Bucksburn to within a mile
of Old Aberdeen.   After heavy rains or a spring thaw
the lower reaches of the river, especially from Kintore
downwards, are apt to be flooded, and in spite of embankments which have been erected along the river's course,
few years pass without serious damage being done to the

Fir Trees at Braemar

crops in low-lying fields. Some parts of Donside scenery, notably at Monymusk (called Paradise), and at Seaton House just below the Cathedral of Old Aberdeen, and before the river passes through the single Gothic arch of the ancient and historical bridge of Balgownie, are very fine—wooded and picturesque, and beloved of more than one famous artist.

The next river is the Ythan, which, rising in the low hills of the Culsalmond district and flowing through the parish of Auchterless and past the charming hamlet of Fyvie, creeps somewhat sluggishly through Methlick and Lord Aberdeen's estates to Ellon. A few miles below Ellon it forms a large tidal estuary four miles in length— a notable haunt of sea-trout, the most notable on the east coast. The river is only 37 miles long. It is slow and winding with deep pools and few rushing streams; moreover its waters have never the clear, sparkling quality of the silvery Dee. Yet at Fyvie and at Gight it has picturesque reaches that redeem it from a uniformity of tameness.

The Ugie, a small stream of 20 miles in length, is the only other river worthy of mention. It joins the sea north of the town of Peterhead. In character it closely resembles the Ythan, having the same kind of deep pools and the same sedge-grown banks.

The Deveron is more particularly a Banffshire river, yet in the Huntly district, it and its important tributary the Bogie (which gives its name to the well-known historic region called Strathbogie) are wholly in Aberdeenshire. The Deveron partakes of the character of the Dee

The Don, looking towards St Machar Cathedral

and the character of the Don.   It is neither so sparkling
and rapid as the one nor so slow and muddy as the other.
Around Huntly and in the locality of Turriff and Eden,
where it is the boundary between the counties, it has
some charmingly beautiful reaches.   Along its banks is

Brig o' Balgownie, Aberdeen

a succession of stately manor-houses, embosomed in trees,
and these highly embellished demesnes enhance its natural
charms.

Lakes are few in Aberdeenshire, and such as exist are
not specially remarkable.   The most interesting historically
are the Deeside Lochs Kinnord and Davan which are

Loch Muick, near Ballater

held by antiquarians to be the seat of an ancient city
Devana—the town of the two lakes. In pre-historic
times there dwelt on the shores of these lakes as also in
the valleys that converge upon them a tribe of people
who built forts, and lake retreats, made oak canoes, and
by means of palisades of the same material created artificial islands. The canoes which have been recovered
from the bed of the loch are hollowed logs thirty feet in
length. Other relics—a bronze vessel and a bronze spear-
head, together with many beams of oak—have been fished
up, all proving the existence of an early Pictish settlement.

Besides these, there is in the same district—but south-
east of Loch-na-gar, another and larger lake called Loch
Muick. From it flows the small river Muick—a tributary
of the Dee, which it joins above Ballater. South-west of
Loch-na-gar is Loch Callater, which drains into the Clunie,
another Dee tributary, which joins the main river at Castleton of Braemar. On the lower reaches of the Dee are
the Loch of Park or Drum, and the Loch of Skene, both
of which drain into the Dee. Both are much frequented
by water-fowl of various kinds.

The Loch of Strathbeg, which lies on the east coast
not far from Rattray Head, is a brackish loch of some
interest. Two hundred years ago, we are told, it was
in direct communication with the sea and small vessels
were able to enter it. In a single night a furious easterly
gale blew away a sand-hill between the Castle-hill of
Rattray and the sea, with the result that the wind-driven
sand formed a sand-bar where formerly there was a clear

Loch Callater, Braemar

water-way.   Since that day the loch has been land-locked and though still slightly brackish may be regarded as an inland loch.

6. Geology.

Geology is the study of the rocks or the substances of which the earthy crust of a district is composed. Rocks are of two sorts: (1) those due to the action of heat, called igneous, (2) those formed and deposited by water, called aqueous.   When the earth was a molten ball, it cooled at the surface, but every now and again liquid portions were ejected from cracks and weak places. The same process is seen in the eruptions of Mount Vesuvius, which sends out streams of liquid lava that gradually cools and forms hard rock.   Such are *igneous* rocks.   But all the forces of nature are constantly at work disintegrating the solid land; frost, rain, the action of rivers and the atmosphere wear down the rocks; and the tiny particles are carried during floods to the sea, where they are deposited as mud or sand-beds laid flat one on the top of the other like sheets of paper.   These are *aqueous* rocks.   The layers are afterwards apt to be tilted up on end or at various angles owing to the contortions of the earth's crust, through pressure in particular directions.   When so tilted they may rise above water and immediately the same process that made them now begins to unmake them.   They too may in time be so worn away that only fragments of them are left whereby we may interpret their history.

Loch of Skene

To these may be added a third kind of rock called *metamorphic*, or rocks so altered by the heat and pressure of other rocks intruding upon them, that they lose their original character and become metamorphosed. They may be either sedimentary, laid down originally by water, or they may be igneous, but in both cases they are entirely changed or modified in appearance and structure by the treatment they have suffered.

The geology of Aberdeenshire is almost entirely concerned with igneous and metamorphic rocks. The whole back-bone of the county is granite which has to some extent been rubbed smooth by glacial action; but in a great part of the county the granite gives place to metamorphic rocks, gneiss, schist, and quartzite. A young geologist viewing a deep cutting in the soil about Aberdeen finds that the material consists of layers of sand, gravel, clay, which are loosely piled together all the way down to the solid granite. This is the glacial drift, or boulder clay, a much later formation than the granite and a legacy of what is called the great Ice Age. Far back in a time before the dawn of history all the north-east of Scotland was buried deep under a vast snow-sheet. The snow consolidated into glaciers just as in Switzerland to-day, and the glaciers thus formed worked their way down the valleys, carrying a great quantity of loose material along with them. When a warmer time came, the ice melted and all the sand and boulders mixed up in the ice were liberated and sank as loose deposits on the land. This is the boulder clay which in and around Aberdeen is the usual subsoil. It consists of rough, half-rounded pebbles, large and small, of clay, sand, and shingle, and makes a very cold and unkindly soil, being difficult to drain properly and slow to take in warmth.

Below this boulder clay are the fundamental rocks. At Aberdeen these are pure granite; but in other parts of the county they are, as we have said, metamorphic, that is, they have been altered by powerful forces, heat and pressure. Whether they were originally sedimentary, before they were altered, is doubtful; some geologists think the crystalline rocks round Fraserburgh and Peterhead were aqueous. Mormond Hill was once a sandstone, and the schists of Cruden Bay and Collieston were clay. The same beds traced to the south are found to pass gradually into sedimentary rocks that are little altered. Whether they were aqueous or igneous originally, they have to-day lost all their original character. No fossils are found in them. These rocks are the oldest and lowest in Aberdeenshire. After their formation, they were invaded from below by intrusive masses of molten igneous rock, which in many parts of the county is now near the surface. This is the granite already referred to. Its presence throughout the county has materially influenced the character and the industry of the people.

Wherever granite enters, it tears its irregular way through the opposing rocks, and sends veins through cracks where such occur. The result of its forcible entrance in a molten condition is that the contiguous rocks are melted, blistered, and baked by the intrusive matter. Why granite should differ from the lava we

see exuding from active volcanoes is explained by the
fact that it is formed deep below the surface where there
is no outlet for its gases. It cools slowly and under great
pressure and this gives it its special character. If found,
therefore, at the surface, as it is in Aberdeen, this is because the rocks once high above it, concealing its presence
have been worn away, which gives some idea of the great
age of the district. One large granitic mass is at Peterhead, where it covers an area of 46 square miles, and forms
the rocky coast for eight miles; but the whole valley of
the Dee as far as Ben Macdhui, and great part of Donside, consist of this intrusive granite. It varies in colour
and quality, being in some districts reddish in tint as at
Sterling Hill near Peterhead, at Hill of Fare, and Coren-
nie; in other parts it is light grey in various shades.

The succession in the order of sedimentary rocks is
definitely settled, and although this has little application
to Aberdeenshire, an outline may be given. The oldest
are the Palaeozoic which includes—in order of age—

> Cambrian,
> Silurian,
> Old Red Sandstone or Devonian,
> Carboniferous,
> Permian.

Of these the only one represented in Aberdeenshire is
the Old Red Sandstone, which occupies a considerable
strip on the coast from Aberdour to Gardenstown, and
runs inland to Fyvie and Auchterless and even as far as
Kildrummy and Auchindoir. The deposit is 1300 feet

thick. A visitor to the town of Turriff is struck by the red colour of many of the houses there, a most unusual variant upon the blue-grey whinstone of the surrounding districts. The explanation is that a convenient quarry of Old Red Sandstone exists between Turriff and Cuminestown. Kildrummy Castle, one of the finest and most ancient ruins in the county, is not like the majority of the old castles built of granite but of a sandstone in the vicinity. The same band extends across country to Auchindoir, where it is still quarried.

The next geological group of Rocks, the Secondary or Mesozoic, includes—in order of age—

> Triassic,
> Jurassic,
> Cretaceous.

These are not at all or but barely represented. A patch of clay at Plaidy, which was laid bare in cutting the railway track, belongs to the Jurassic system and contains ammonites and other fossils characteristic of that period. Over a ridge of high ground stretching from Sterling Hill south-eastwards are found numbers of rolled flints belonging to the Cretaceous or chalk period, but the probability is that they have been transported from elsewhere by moving ice and are not in their natural place.

The Tertiary epoch is just as meagrely represented as the Secondary. Yet this is the period which in other parts of the world possesses records of the most ample kind. The Alps, the Caucasus, the Himalayas were all upheaved in Tertiary times; but of any corresponding

activity in the north-east of Scotland, there is no trace.
It is only when the Tertiary merges in the Quaternary
period that the history is resumed. The deposits of the
Ice Age, when Scotland was under the grip of an arctic
climate, are much in evidence all over the county and
have already been referred to. It is necessary to treat
the subject in some detail.

During the glacial period, the snow and ice accu-
mulated on the west side of the country, and overflowed
into Aberdeenshire. There were several invasions owing
to the recurrence of periods of more genial temperature
when the ice-sheet dwindled. One of the earlier inroads
probably brought with it the chalk flints now found west
of Buchan Ness; another brought boulders from the district of Moray. South of Peterhead a drift of a different
character took place. Most of Slains and Cruden as well
as Ellon, Foveran, and Belhelvie are covered with a reddish clay with round red pebbles like those of the Old
Red Sandstone. This points to an invasion of the ice-
sheet from Kincardine, where such deposits are rife.
Dark blue clay came from the west, red clay from the
south, and in some parts they met and intermixed as at
St Fergus. A probable third source of glacial remains
is Scandinavia. In the Ice Age Britain was part of the
continental mainland, the shallow North Sea having been
formed at a subsequent period. The low-lying land at
the north-east of the county was the hollow to which the
glaciers gravitated from west and south and east, leaving
their *débris* on the surface when the ice disappeared. So
much is this a feature of Buchan that one well-known

geologist has humorously described it as the riddling heap of creation.

Both the red and the blue clay are often buried under the coarse earthy matter and rough stones that formed the residuum of the last sheet of ice. This has greatly increased the difficulty of clearing the land for cultivation. Moreover a clay sub-soil of this kind, which forms a hard bottom pan that water cannot percolate through, is not conducive to successful farming. Drainage is difficult but absolutely necessary before good crops will be produced. Both difficulties have been successfully overcome by the Aberdeenshire agriculturist, but only by dint of great expenditure of time and labour and money.

The district of the clays is associated with peat beds. There is peat, or rather there was once peat all over Aberdeenshire, but the depth and extent of the beds are greatest where the clay bottom exists. A climate that is moist without being too cold favours the growth of peat and the Buchan district, projecting so far into the North Sea and being subject to somewhat less sunshine than other parts of the county, provides the favouring conditions. The rainfall is only moderate but it is distributed at frequent intervals, and the clay bottom helps to retain the moisture and thus promotes the growth of those mosses which after many years become beds of peat. These peat beds for long provided the fuel of the population. In recent years they are all but exhausted, and the facility with which coals are transported by sea and by rail is gradually putting an end to the "casting" and drying of peats.

Moraines of rough gravel—the wreckage of dwindling glaciers—are found in various parts of the Dee valley. The soil of Deeside has little intermixture of clay and is thin and highly porous. It follows that in a dry season the crops are short and meagre. The Scots fir, however, is partial to such a soil, and its ready growth helps with the aid of the natural birches to embellish the Deeside landscape.

In the Cairngorms brown and yellow varieties of quartz called "cairngorms" are found either embedded in cavities of the granite or in the *detritus* that accumulates from the decomposition of exposed rocks. The stones, which are really crystals, are much prized for jewellery, and are of various colours, pale yellow (citrine), brown or smoky, and black and almost opaque. When well cut and set in silver, either as brooches or as an adornment to the handles of dirks, they have a brilliant effect. Time was when they were systematically dug and searched for, and certain persons made a living by their finds on the hill-sides; but now they are more rare and come upon only by accident.

7. Natural History.

As we have seen in dealing with the glacial movements, Britain was at one time part of the continent and there was no North Sea. At the best it is a shallow sea, and a very trifling elevation of its floor would re-connect Scotland with Europe. It follows that our country was inhabited by the same kind of animals as inhabited Western Europe. Many of them are now extinct, cave-bears, hyaenas and sabre-toothed tigers. All these were starved out of existence by the inroads of the ice. After the ice disappeared this country remained joined to the continent, and as long as the connection was maintained the land-animals of Europe were able to cross over and occupy the ground; if the connection had not been severed, there would have been no difference between our fauna and the animals of Northern France and Belgium. But the land sank, and the North Sea filled up the hollow, creating a barrier before all the species in Northern Europe had been able to effect a footing in our country. This applies both to plants and animals. While Germany has nearly ninety species of land animals, Great Britain has barely forty. All the mammals, reptiles and amphibians that we have, are found on the continent besides a great many that we do not possess. Still Scotland can boast of its red grouse, which is not seen on the continent.

With every variety of situation, from exposed sea-board to sheltered valley and lofty mountain, the flora of Aberdeenshire shows a pleasing and interesting variety. The plants of the sea-shore, of the waysides, of the river-banks, and of the lowland peat-mosses are necessarily different in many respects from those of the great mountain heights. It is impossible here to do more than indicate one or two of the leading features. The sandy tracts north of the Ythan mouth have characteristic plants, wild rue, sea-thrift, rock-rose, grass of Parnassus,

catch-fly (*Silene maritima*). The waysides are brilliant
with blue-bells, speedwell, thistles, yarrow and violas.
The peat-mosses show patches of louse-wort, sundew,
St John's wort, cotton-grass, butterwort and ragged
robin. The pine-woods display an undergrowth of blae-
berries, galliums, winter-green, veronicas and geraniums.
The *Linnaea borealis* is exceedingly rare, but has a few
localities known to enterprising botanists. The whin and
the broom in May and June add conspicuous colouring
to the landscape while a different tint of yellow shines
in the oat-fields, which are throughout the county more
or less crowded with wild mustard or charlock. The
granitic hills are all mantled with heather (common ling,
Calluna erica) up to 3000 feet, brown in winter and
spring but taking on a rich purple hue when it breaks
into flower in early August. The purple bell-heather
does not rise beyond 2000 feet and flowers much earlier.
Through the heather trails the stag-moss, and the pyrola
and the genista thrust their blossoms above the sea of
purple. The cranberry, the crow-berry and the whortle-
berry, and more rarely the cloudberry or Avron (*Rubus
chamaemorus*) are found on all the Cairngorms. The
Alpine rock-cress is there also, as well as the mountain
violet (*Viola lutea*), which takes the place of the hearts-
ease of the lowlands. The moss-campion spreads its
cushions on the highest mountains; saxifrages of various
species haunt every moist spot of the hill-sides and the
Alpine lady's mantle, the Alpine scurvy-grass, the Alpine
speedwell, the trailing azalea, the dwarf cornel (*Cornus
suecica*), and many other varieties are to be found by
those who care to look for them.

As we have said, no trees thrive near the coast. The easterly and northerly winds make their growth precarious, and where they have been planted they look as if shorn with a mighty scythe, so decisive is the slope of their branches away from the direction of the cold blasts. Their growth too in thickness of bole is painfully slow, even a period of twenty years making no appreciable addition to the circumference of the stem. Convincing evidence exists that in ancient times the county was closely wooded. In peat-bogs are found the root-stems of Scots fir and oak trees of much larger bulk than we are familiar with now. The resinous roots of the fir trees, dug up and split into long strips, were the fir-candles of a century ago, the only artificial light of the time.

The district is not exceptional or peculiar in its fauna. The grey or brown rat, which has entirely displaced the smaller black rat, is very common and proves destructive to farm crops—a result partially due to the eradication of birds of prey, as well as of stoats and weasels, by gamekeepers in the interest of game. The prolific rabbit is in certain districts far too numerous and plays havoc with the farmer's turnips and other growing crops. Brown hares are fairly plentiful but less numerous than they were in the days of their protection. Every farmer has now the right to kill ground game (hares and rabbits) on his farm and this helps to keep the stock low. The white or Alpine hare is plentiful in the hilly tracts and is shot along with the grouse on the grouse moors. The otter is occasionally trapped on the rivers, and a few foxes are shot on the hills. The mole is in evidence everywhere up to the 1500 feet level, by the mole-heaps he leaves in every field, and the mole-catcher is a familiar character in most parishes. The squirrel has worked his way north

Deer in time of snow

during the last sixty years, and is now to be found in every fir-wood. The graceful roedeer is also a denizen of the pine-woods, whence he makes forays on the oat-fields. The red-deer is abundant on the higher and more remote hills, and deer-stalking is perhaps the most exciting as it certainly is the most exacting of all forms of Scottish sport.   The pole-cat is rarely seen; he is best known to the present generation in the half-domesticated breed called the ferret.   The hedge-hog, the common shrew, and the water-vole are all common.

The birds are numerous and full of interest.   The coast is frequented by vast flocks of sea-gulls, guillemots,

The Dunbuy Rock

and cormorants, while the estuary of the Ythan has many visitants such as the ringed plover, the eider-duck, the shelduck, the oyster-catcher, redshank, and tern.   On the north bank of this river the triangular area of sand-dunes between Newburgh and Collieston is a favourite nesting-place for eider-duck and terns.   The nests of the eider-duck, with their five large olive-green eggs embedded in the soft down drawn from the mother's breast, are found in great numbers amongst the grassy bents. The eggs of the tern, on the other hand, are laid in a mere hollow of the open sand, but so numerous are they that it is almost impossible for a pedestrian to avoid treading upon them. Puffins or sea-parrots are conspicuous amongst the many sea-birds that frequent Dunbuy Rock. This island rock, half-way up the eastern coast, is a typical sea-bird haunt, where gulls, puffins, razorbills and guillemots are to be seen in a state of restless activity. A colony of black-headed gulls has for a number of years bred and multiplied in a small loch near Kintore. A vast number of migratory birds strike the shores of Aberdeenshire every year in their westward flight. The waxwing, the hoopoe, and the ruff are occasional visitors, the great northern diver and the snow-bunting being more frequent.

The game-birds of the district are the partridge and the pheasant in the agricultural region, and the red grouse on the moors. The higher hills, such as Loch-na-gar, have ptarmigan, while the wooded areas bordering on the highland line are frequented by black-cock and capercailzie. These last are a re-introduction of recent years and seem to be multiplying; but, like the squirrel, they are destructive to the growing shoots of the pine trees and are not encouraged by some proprietors. The lapwing or green plover's wail is an unfailing sound throughout the county in the spring. These useful birds are said to be fewer than they were fifty years ago—a result probably due to

the demand for their eggs as a table delicacy. After the
first of April it is illegal to take the eggs, and this partial
protection serves to maintain the stock in fair numbers.
The starling, which, like the squirrel, was unknown in
this district sixty years ago, has increased so rapidly that
flocks of them containing many thousands are now a
common sight in the autumn. The kingfisher is met
with, very, very rarely on the river-bank, but the dipper
is never absent from the boulder-strewn beds of the
streams. The plaintive note of the curlew and the
shriller whistle of the golden plover break the silence
of the lonely moors. The golden eagle nests in the
solitudes of the mountains and may occasionally be seen,
soaring high in the vicinity of his eyrie.

Of fresh-water fishes, the yellow or brown trout is
plentiful in all the rivers, especially in the Don and the
Ythan. The migratory sea-trout and the salmon are also
caught in each, although the Dee is pre-eminently the
most productive. The salmon fisheries round the coast
and at the mouth of the rivers are a source of considerable revenue. The fish are caught by three species of
net, bag-nets (floating nets) and stake-nets (fixed) in the
sea, and by drag-nets or sweep-nets in the tidal reaches
of the rivers. Time was when drag-nets plied as far
inland as Banchory-Ternan (19 miles), but these have
gradually been withdrawn and are now relegated to a
short distance from the river mouth, the rights having
been bought up by the riparian proprietors further up
the river, who wish to obtain improved opportunities
for successful angling. The Dee has, in this way, been

so improved that it is now perhaps the finest salmon-angling river in Scotland.

The insects of the district call for little remark. Butterflies are few in species and without variety. It is only in certain warm autumns that the red admiral puts in an appearance. The cabbage-white, the tortoise-shell, and an occasional meadow-brown and fritillary are the prevailing species.

The waters of the Ythan, the Ugie, and the Don are frequented by fresh-water mussels which produce pearls. These grow best on a pebbly bottom not too deep and are 3 to 7 inches long and $1\frac{1}{2}$ to $2\frac{1}{2}$ broad. The internal surface is bluish or with a shade of pink. The search for these mussels in order to secure the pearls they may and do sometimes contain was once a recognised industry. To-day it is spasmodic and mostly in the hands of vagrants. Many beds are destroyed before the mussels are mature and this lessens the chances of success. The pearl-fisher usually wades in the river, making observation of the bottom by means of a floating glass which removes the disturbing effect of the surface ripple. He thus obtains a clear view of the river-bed, and by means of a forked stick dislodges the mussels and brings them to bank, 150 making a good day's work. He opens them at leisure and finds that the great majority of his pile are without pearls. If he be lucky enough, however, to come upon a batch of mature shells he may find a pearl worth £20. As a rule the price is not above ten or twenty shillings. Much depends on the size and the colouring. The most valuable are those of a pinkish hue.

8. Round the Coast.

The harbour-mouth, which is also the mouth of the Dee, is the beginning of the county on the sea-board. It is protected by two breakwaters, north and south, which shelter the entrance channel from the fury of easterly and north-easterly gales. To the south, in Kincardineshire, is the Girdleness lighthouse, 185 feet high, flashing a light every twenty seconds with a range of visibility stated at 19 miles. To the north of the harbour entrance are the links and the bathing station. The latter was erected in 1895 and has since been extended, every effort being made to add to the attractiveness of the beach as a recreation ground. A promenade, which will ultimately extend to Donmouth, is in great part complete; and all the other usual concomitants of a watering-place have been introduced with promising success so far, and likely to be greater in the near future.

From Donmouth the northward coast presents little of interest. All the way to the estuary of the Ythan is a region of sand-dunes bound together by marum grass and stunted whins, excellent for golf courses, but lacking in variety. In the sandy mounds in the vicinity of the Ythan have been found many flint chippings and amongst them leaf-shaped flint arrow-heads, chisels and cores, as well as the water-worn stones on which these implements were fashioned. These records of primitive man as he was in the later Stone Age are conspicuous here, and are

Girdleness Lighthouse

to be seen in other parts of the county. In the rabbit burrows, which are abundant in the dunes, the stock-dove rears her young. In 1888 a migratory flock of sand-grouse took possession of the dunes, and remained for one season.

Beyond the Ythan are the Forvie sands—a region of hummocks under which a whole parish is buried. The destruction of the parish took place several centuries ago, when a succession of north-easterly gales, continued for many days, whipped up the loose sand of the coast-dunes and blew it onward in clouds till the whole parish, including several valuable farms, was entirely submerged. The scanty ruins of the old church of Forvie is the only trace left of this sand-smothered hamlet.

Not far from the site of the Forvie church is a beautiful semi-lunar bay called Hackley Bay, where for the first time since Aberdeen was left behind, rocks appear, hornblende, slate, and gneiss. At Collieston, a village consisting of a medley of irregularly located cottages scrambling up the cliff sides, a thriving industry used to be practised, the making of Collieston "speldings." These were small whitings, split, salted and dried on the rocks. Thirty years ago they were considered something of a delicacy and were disposed of in great quantities; now they have lost favour and are seldom to be had. At the north end of the village is St Catherine's Dub, a deep pool between rocks, on which one of the ships of the Spanish Armada was wrecked in 1588. Two of the St Catherine's cannon very much corroded have been brought up from the sea-floor. One of them is still to be seen at Haddo House, the seat of the Earl of Aberdeen.

Northward we come upon a region of steep grassy braes, consisting of soft, loamy clay, 20 to 40 feet deep, and covered with luxuriant grasses in summer and ablaze with golden cowslips in the spring months. Along the coast are several villages which once populous with busy and hardy fishermen are now all but tenantless. Such are Slains and Whinnyfold crushed out of activity by the rise of the trawling industry. The next place of note is Cruden Bay Hotel built by the Great North of Scotland Railway Company, and intended to minister specially to the devotees of golf, for which the coast links are here eminently suitable. The fine granite building facing the sea is a conspicuous landmark. Just north of the Hotel is the thriving little town of Port Errol, through which runs the Cruden burn—a stream where sea-trout are plentifully caught at certain seasons. The next promi-nent object is Slains castle—the family seat of the Earls of Errol. It stands high and windy, presenting a bold front to the North Sea breezes. All its windows on the sea-face are duplicate, a necessary precaution in view of the fierceness of the easterly gales. Very few plants grow in this exposed locality, and these only in the hollow and sheltered ground behind the castle, where some stunted trees and a few garden flowers struggle along in a pre-carious existence. As we proceed, the rocky coast rises higher and bolder and presents variable forms of great beauty. Beetling crags enclose circular bays with per-pendicular walls on which the kittiwake, the guillemot,

Sand Hills at Cruden Bay

the jackdaw and the starling breed by the thousand. The rock of Dunbuy, a huge mass of granite, surrounded by the sea, and forming a grand rugged arch, is a summer haunt of sea-birds and rock-pigeons.

After this, we reach the picturesque and much visited Bullers of Buchan—a wide semi-circular sea cauldron, the sides of which are perpendicular cliffs. The pool has no entry except from the seaward side, and it is only in calm weather that a boat is safe to pass through the low, open archway in the cliff. In rough weather, the waves rush through the narrow archway with terrific force, sending clouds of spray far beyond the height of the cliffs. Under proper conditions the scene is one of the grandest in Aberdeenshire, and is a fitting contrast to the sublimely impressive scenes at the source of the Dee, right at the other end of the county. Beyond the Bullers, the coast consists of high granite rocks, behind which are wind-swept moors. Near Boddam is Sterling Hill quarry, the source of the red-hued Peterhead granite. Here too is Buchan Ness, the most easterly point on the Scottish coast, and a fitting place for a prominent lighthouse. The lantern of the circular tower (erected in 1827) stands 130 feet above high-water mark and flashes a white light once every five seconds. The light is visible at a distance of 16 nautical miles.

At Peterhead, which is a prosperous fishing centre and the eastern terminus of the bifurcate Buchan line of rail-way, is a great convict prison, occupying an extensive range of buildings on the south side of the Peterhead bay. The convicts are employed in building a harbour of

"The Pot," Bullers o' Buchan

refuge, which is being erected under the superintendence of the Admiralty at a cost of a million of money. The coast onwards to the Ugie mouth is still rocky, but from the river to Rattray Head, the rocks give place to sand-dunes similar in character to those further south. Alongside of the dunes is a raised sea beach. They form the links of St Fergus. Rattray Head is a rather low reef of rock running far out to sea and highly suitable as a lighthouse station. In the course of twelve years, the reef was responsible for 24 shipwrecks. The lighthouse erected in 1895 is 120 feet high and the light gives three flashes in quick succession every 30 seconds. It is visible 18 miles out to sea. Beyond this point is a region of bleak and desolate sands. Not a tree nor a shrub is to be seen. The inland parts are under cultivation, but the general aspect of the country is dismal and dreary, and the very hedgerows far from the sea-board lean landwards as if cowering from the scourges of the north wind's whip. The country is undulatory without any conspicuous hill. Beyond Rattray Head is the Loch of Strathbeg already referred to. The tradition goes that the same gale as blighted Forvie silted up this loch and contracted its connection with the sea. On the left safely sheltered from the sea-breezes are Crimonmogate, Cairness and Philorth—all mansion-houses surrounded by wooded grounds. At the sea-edge stand St Combs (an echo of St Columba), Cairnbulg and Inverallochy. Here occurs another raised sea beach. Our course from Rattray Head has been north-west and thus we reach the last important town on the coast—Fraserburgh.

Buchan Ness Lighthouse

Fraserburgh lies to the west of its bay. Founded by one of the Frasers of Philorth (now represented by Lord Saltoun), it is like Peterhead a thriving town. Like Peter-

Kinnaird Lighthouse, Fraserburgh

head too, it is the terminus of one fork ot the Buchan Railway and a busy fishing centre. In the month of July, which is the height of the herring season, " the Broch," as it is called locally, is astir with life from early morn.

More herrings are handled at Fraserburgh than anywhere else on this coast, from Eyemouth to Wick. Between Fraserburgh and Broadsea is Kinnaird's Head. Here we have another lighthouse which has served that purpose for more than a century, an old castle having been converted to this use in 1787. It was one of the first three lighthouses in Scotland. Kinnaird's Head is believed to

Entrance to Lord Pitsligo's Cave, Rosehearty

be the promontory of the Taixali mentioned by the Alexandrian geographer Ptolemy as being at the entrance of the Moray Firth. Here the rocks are of moderate height but further west they fall to sea-level and continue so past Sandhaven and Pitullie to Rosehearty. A low rocky coast carries us to Aberdour bay, where beds of Old Red Sandstone and conglomerate rise to an altitude of 300 feet.

Aberdour Shore, looking N.W.

The conglomerate extends to the Red Head of Pennan—once a quarry for mill-stones—where an attractive and picturesque little village nestles at the base of the cliff. The peregrine falcon breeds on the rocky fastnesses of these lofty cliffs, which continue to grow in height and grandeur till they reach their maximum (400 feet) at Troup Head. Troup Head makes a bold beginning for the county of Banff.

9. Weather and Climate. Temperature. Rainfall. Winds.

The climate of a county depends on a good many things, its latitude, its height above sea-level, its proximity to the sea, the prevailing winds, and especially as regards Scotland whether it is situated on the east coast or on the west. The latitude of Great Britain if the country were not surrounded by the sea would entitle it to a temperature only comparable to that of Greenland but its proximity to the Atlantic redeems it from such a fate. The Atlantic is 3° warmer than the air and the fact that the prevailing winds are westerly or south-westerly helps to raise the mean temperature of the western counties higher than that of those on the east. The North Sea is only 1° warmer than the air so that its influence is less marked.

Still, considering its latitude (57°—57° 40′), Aberdeenshire enjoys a comparatively moderate climate. It is neither very rigorous in winter nor very warm in

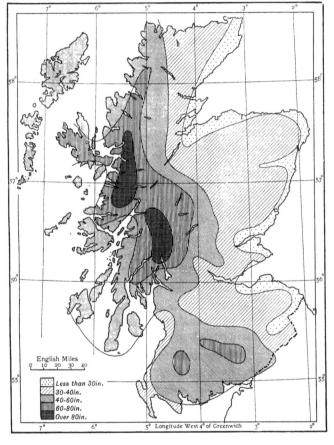

Rainfall Map of Scotland. (After Dr H. R. Mill)

summer. Of course in a large county a distinction must be drawn between the coast temperature and that of the high lying districts such as Braemar. The fringe round the coast is in the summer less warm than the inland parts, a result due to the coolness of the enclosing sea, but in the winter this state of affairs is reversed and the uplands are held in the grip of a hard frost while the coast-side has little or none.

The mean temperature of Scotland is 47°, while Aberdeen has 46°·4 and Peterhead 46°·8. That of Braemar, the most westerly station in the county, though in reality very little lower, is arrived at by entirely different figures; the temperature being much higher during July, August and September, but lower in December, January and February. Braemar is 1114 feet above sea-level and since there is a regular and uniform decline in temperature to the extent of 1° for every 270 feet above the sea, the temperature of this hill-station should be low. As a matter of fact, from June to September it is only 9° and in October 7°·5 below that of London. Yet its maximum is 10° higher than is recorded at Aberdeen, only in winter its minimum is 20° lower than the minimum of the coast.

Braemar and Peterhead as lying at the two extremes of the county may be compared. Peterhead receives the uninterrupted sweep of the easterly breezes, for it has no shelter or protection either of forests or mountains. The impression a visitor takes is that Peterhead is an exceptionally cold place. As a fact, its mean winter temperature is above the average for Scotland, but the lack of shelter and the constant motion of the air give an impression of

Inverey near Braemar

5—2

coldness. In the summer and autumn its mean falls below that of Scotland. It is therefore less cold in the cold months and less warm in the warm months than Braemar and has a seasonal variation of only 16°·3 between winter and summer, whereas Edinburgh has a range of 21° and London of 26°.

The rainfall over the whole county is also moderate, ranging from less than 25 inches at Peterhead—the driest part of the area—to 40 inches at Braemar, and 32 at Aberdeen. This is a small rainfall compared with 60 or 70 inches on parts of the west coast. The driest months in Aberdeenshire are April and May, and generally speaking less rain falls in the early half of the year when the temperature is rising than in the later half when the temperature is on the decline. Two inches is about the average for each month from February to June, but October, November and December are each over three inches. The most of the rainfall of Scotland comes from the west and south. This explains why the west coast is so much wetter than the east. The westerly winds from the Atlantic, laden with moisture, strike upon the high lands of the west, but exhaust themselves before they reach the watershed and, having precipitated their moisture between that and the coast, they reach the east coast comparatively dry. Braemar just under the watershed is relatively dry. Its situation as an elevated valley, 1114 feet above sea-level and surrounded on three sides by hills of from three to four thousand feet, and the fact that it is 60 miles from the sea combine to make it one of the most bracing places and give it one of the finest summer climates in the British Isles.

This sufficiently accounts for its popularity as a health resort. May is its driest month, October its wettest.

Easterly winds bring rain to the coast, but as a rule the rain extends no further inland than 20 miles. Easterly winds prevail during March, April and May, which make this season the most trying part of the year for weakly people. In summer the winds are often northerly, but the prevailing winds of the year, active for 37 per cent. of the 365 days or little less than half, are west and south-west. East winds bring fog, and this is most prevalent in the early summer, June being perhaps the worst month. The greatest drawback to the climate from an agriculturist's point of view is the lateness of the spring. The summer being short, a late spring means a late harvest, which is invariably unsatisfactory.

The low rainfall of the county is favourable to sunshine. Aberdeen has 1400 hours of sunshine during the year in spite of fogs and east winds; the more inland parts being beyond the reach of sea-fog have an even better record.

The great objection—an objection taken by folks who have spent part of their life in South Africa or Canada—is the variableness of the climate from day to day. There is not here any fixity for continued periods of weather such as obtains in these countries. The chief factor in this variability is our insular position on the eastern side of the Atlantic. When, on rare occasions, as sometimes happens in June or in September, the atmosphere is settled, Aberdeenshire enjoys for a few weeks weather of the most salubrious and delightful kind.

10.   The   People — Race,   Language, Population.

The blood of the people of Aberdeenshire, though in the main Teutonic, has combined with Celtic and other elements, and has evolved a distinctive type, somewhat different in appearance and character from what is found in other parts of Scotland.   How this amalgamation came about must be explained at some length.

The earliest inhabitants of Britain must have crossed from Europe when as yet there was no dividing North Sea.   They used rough stone weapons (see p. 114) and were hunters living upon the products of the chase, the mammoth, reindeer and other animals that roamed the country.   Such were *palaeolithic* (ancient stone) men. Perhaps they never reached Scotland : at least there is no trace of them in Aberdeenshire.   They were followed by *neolithic* (new stone) men, who used more delicately carved weapons, stone axes, and flint arrows.   Traces of these are to be found in Aberdeenshire.   A few short cists containing skeletal remains have been found in various parts of the county.   In the last forty years some fifteen of these have been unearthed.   From these anthropologists conclude that neolithic men lived here at the end of the stone age, men of a muscular type, of short stature and with broad short faces.   They were mighty hunters hunting the wild ox, the wolf and the bear in the dense forests which, after the Ice Age passed, overspread the north-east.   They clothed themselves against

the cold in' the skins of the animals which they made
their prey and were a rude, savage, hardy race toughened
by their mode of life and their fierce struggle for existence.   They did not live by hunting alone; they possessed
herds of cattle, swine and sheep and cultivated the ground
but probably only to a slight extent.   Their weapons
were rude arrow-heads, flint knives and flint axes; and a
considerable number of these primitive weapons as well
as the bones of red-deer and the primeval ox—*bos primi-
genius*—have been recovered from peat-mosses and elsewhere throughout the district.   Such have been found at
Barra, at Inverurie and at Alford.

Besides these remains, have been found urns made of
boulder clay, burned by fire and rudely ornamented.
These were very likely their original drinking vessels,
afterwards somewhat modified as food vessels, and were,
it is supposed, deposited in graves with a religious motive
in accordance with the belief common among primitive
peoples that paradise is a happy hunting-ground in which
the activities of the present life will continue under more
favourable conditions.

In addition to these relics the county has a great
number of stone circles, circles of large upright boulders
set up not at hap-hazard but evidently with some definite
object in view.   These will be dealt with in a later
chapter.   The probability is that the so-called Pictish
houses, the earth or Eirde houses found on Donside and
the lake dwellings at Kinnord already referred to, were the
homes of these people.   But the whole subject is by no
means clear.   The general opinion is that the north-east

was first inhabited by Picts, who may or may not have been Iberians, and that after the Picts came the Celts; but some critics hold that the Picts were only earlier Celts. In any case the Stone Age was succeeded by the Bronze Age, when Bronze took the place of Stone in the formation of weapons. The Celts made their way through Central France to Britain and ultimately to Scotland. Unlike the people they found in possession of Scotland, they were tall (5 ft. 9 in.). These are the ancestors of the Gaelic speaking people of Scotland. They are supposed to have amalgamated to some extent with the Neolithic men whom they found on the spot, and it is certain that they were christianised at an early period. Later on Teutonic tribes, tall, longheaded and fairhaired men crossed from the Baltic to Britain and in due course they too reached Aberdeenshire. But up to the time of David I (1124–1153) the population and institutions of the north-east were entirely Celtic. The Saxon or Teutonic element was introduced by way of the coast and the trading towns. From the towns it spread to the country districts. When Henry II expelled the Flemish traders from England many migrated to the north and formed settlements in many parts of the country, establishing trade and handicraft, particularly weaving, and reclaiming waste land. The defeat of Comyn, the Earl of Buchan, by Bruce in 1308, when Bruce harried Buchan from end to end and spared none, opened the way for lowland immigrants and not only gave an impetus to Teutonic settlements, but helped to kill out the Celtic language and the Celtic ways. These immigrants are really the

ancestors of the present Aberdeenshire people, but they
have been greatly modified by absorbing the Celtic population and mixing with it, for though reduced by slaughter,
and by an exodus to the hills, it had not entirely disappeared. Scandinavians from Norway and Denmark also
found a footing at various periods in this north-eastern
region and these elements are all blended in the modern
Aberdonian. Celt, Saxon, Fleming and Scandinavian
came in one after the other and possessed the land, forming a new people in which all these elements were fused.

Aberdonians are credited with a distinct individuality,
partly the result of race, partly due to environment. The
strain of practicality in the Teuton toned down the Celtic
imagination and warmth of feeling, and added a certain
tincture of the phlegmatic such as is so prominent in the
Dutchman. Hence the cautious "canny" nature of the
typical Aberdonian, dreading innovations, resisting agricultural novelties, and disliking ecclesiastical changes.
They have been described as people

> Who are not fond of innovations,
> Nor covet much new reformations;
> They are not for new paths but rather
> Each one jogs after his old father.

This requires some elucidation. They are far from
slothful or indifferent. They will uphold with zeal the
cause they think right, but they must first reach assured
conviction that it *is* right. They are not swift nor slow
to change, but firm.

The Celtic population was in fact absorbed, as we
have said, but a certain contingent betook themselves to

the mountains and for long kept up a warfare of reprisals upon those who had dispossessed them. This caused no end of trouble in Aberdeenshire but not without its uses for it braced the occupants in the arts of defence and made them alert and courageous.

No less potent a factor in the evolution of the Aberdonian has been his struggle with a well-nigh irreclaimable soil. The county is without mineral wealth, and the only outlet for his energy was found in attacking the boulder-strewn moors and in clearing them for the plough. To this he set his mind in the eighteenth century with grim determination. Small farmers and crofters by dint of great personal toil and life-long self-sacrifice transformed stony tracts of poor and apparently worthless land into smiling and productive fields. It is this struggle with a malignant soil, more than anything else, that has made the Aberdonian; one triumph led on to another, and to-day the spirit of enterprise in farming is nowhere more pronounced than in this difficult county.

The place names are almost entirely Celtic, and even when they appear to be Saxon they are only Gaelic mispronounced or assimilated to something better known. The parish of King Edward might very plausibly be referred to the northern visits paid by the Hammer of the Scots, but it is really Kinedar, with the Gaelic *Kin* (seen in Kinnaird, Kintore and Malcolm Canmore), meaning a head.

The county has a distinctive dialect, really imported and originally uniform with the dialect of the Mearns, and of Northumbria, the dialect spoken at one time all the way from Forth to Humber. To-day it is called the Buchan Doric and though varying somewhat in different parts of the county and hardly intelligible in the Highlands of Braemar, where Gaelic still survives, it is a Teutonic speech with a thin tincture of Gaelic words such as *bourach, closach, clachan, brochan,* etc.

The dialect contains many vocables not found in literary English, such as *byous* and *ondeemis* for extraordinary, but where the words are English, they are greatly altered. It is characterised by broad, open vowels; "boots" is pronounced "beets," "cart" is "cairt," "good" is "gweed." The final *l* is dropped; "pull" is "pu," "fall" is "fa." Final *ol* becomes *ow*; "roll" is made "row," and "poll" is "pow." *Wh* is always *f*: "white" is "fite" and "who?" (interrogative) is "fa?" It is rich in diminutives like the Dutch—*a lassie, a basketie.* The finest embodiment of this striking dialect, giving permanent life to its wealth of pathos and expressiveness, is Dr William Alexander's *Johnny Gibb of Gushetneuk.*

Scots wha hae, which is supposed to be a characteristic phrase common to all the dialects, would be in Buchan —*Scots at hiz,* which is largely Norse. "The quynie coudna be ongrutten" is Buchan for "The little girl could not help crying."

The population of the county which a hundred years before was 121,065 in 1901 was 304,439. Since the county contains 1970 square miles this brings out an average of 154 to the square mile—just a little over the average of Scotland as a whole, but as Aberdeen city accounts for more than half of the total, and towns like

Peterhead and Fraserburgh between them represent
25,000, the figure is greatly reduced for the rural districts.
The country districts are but thinly peopled, especially
on the Highland line, and the tendency is for the rural
population to dwindle. They either emigrate to Canada,
which is a regular lodestone for Aberdonians, or they
betake themselves to the towns, chiefly to Aberdeen itself.
Except in and around the principal town, the county has
hardly any industries that employ many hands. Agriculture is the main employment, and modern appliances
enable the farmer to do his work with fewer helps than
formerly: hence the depopulation of the rural districts.
The towns tend to grow, the rural parishes to become
more sparsely inhabited.

11. Agriculture.

This is the mainstay of the county, and considering
the somewhat uncertain climate, the shortness of the
summer and the natural poverty of the soil, it has been
brought to marvellous perfection. The mountainous
regions are necessarily cut off from this industry except in
narrow fringes along the river banks, but in the low-lying
area it is safe to say that every acre of ground worth
reclaiming has been put to the plough. A century ago
the industry was rude and ill-organised, the county being
without roads and without wheeled vehicles, but the
advent of railways gave an impetus to the farming instinct
and an extraordinary activity set in to reclaim waste land

by clearing it of stones, by trenching, by draining and manuring it. The proprietors were usually agreeable to granting a long lease at a nominal rent to any likely and energetic man who was willing to undertake reclamations and take his chance of recouping himself for outlays before his lease expired. Being thus secured, the farmer or crofter had an incentive to put the maximum of labour into his holding. He often built the dwelling-house, and as a rule made the enclosures by means of the stones, which, with great labour, he dragged from the fields. In this way a great acreage was added to the arable land of the county, and though some of it has fallen into pasture since the great boom in agricultural prices during the seventies in last century, the greater part of the reclaimed soil is still in cultivation.

The area of the county, exclusive of water and road-ways, is 1955 square miles, or 1,251,451 acres. Of this exactly one half is under cultivation, 628,523 acres. When we remember that Scotland contains some nineteen million of acres and that only 25 per cent. of this acreage is arable land, it is apparent that Aberdeen with its 50 per cent. is one of the most cultivated areas. As a matter of fact it has by far the largest acreage under cultivation of any Scottish county. Next to it is Perth-shire with 336,251 acres. The uncultivated half is made up of mountain, moor and woodlands. Part of this is used for grazing sheep, as much as 157,955 acres being thus utilised. In the matter of woods and plantations the county with its 105,931 acres stands next to Inver-ness-shire, which has 145,629. The trees grown are mostly larch and pine and spruce, but the deciduous trees, or hard woods, the beech, elm and ash, are not uncommon in the low country, more especially as ornamental trees around the manor-houses of the proprietors.

The crops chiefly cultivated are oats, barley, turnips and potatoes. Wheat is not grown except now and again in an odd field. The climate is too cold, the autumn heat never rising to the point of ripening that crop satisfactorily. Oats is the most frequent crop, and Aberdeenshire is the oat-producing county of Scotland. A fifth of the whole acreage under this crop in Scotland belongs to Aberdeenshire. Perth, which is next, has only one-third of the Aberdeenshire oat-area. Twenty thousand acres are devoted to barley, only one-tenth of the barley-area in Scotland. Over seven thousand acres go to potatoes; the southern counties have a soil better adapted to produce good potatoes; Forfar, Fife, Perth and Ayrshire excel in this respect and all these give a larger acreage to this crop. As regards turnips, however, Aberdeenshire is easily first. Being a great cattle rearing and cattle feeding district, it demands a large tonnage of turnip food. It is estimated that a million and a half tons of turnips are consumed every year in the county.

As regards cattle and horses the county has first place in Scotland. In 1909 there were 204,490 agricultural horses in the country and of these 31,592 were in Aberdeenshire, while of 1,176,165 cattle it had 168,091. It has a quarter of a million sheep, but here it falls behind other counties, notably Argyll, which has nearly a million, or one-seventh of all the sheep in Scotland.

Aberdeenshire is a county of small holdings. No other county has so many tenants. Over five thousand of these farm from five to 50 acres, while there are nearly four thousand who farm areas ranging from 50 to 300. This is part of the secret of its success. Earlier, the number of small farms was greater, the tendency being in the direction of throwing several smaller holdings together to make a large farm.

The industry has been a progressive one. Up to the Union in 1707 tillage was of the most primitive kind. Sheep-farming for the sake of exporting the wool had been the rule, but the Union stopped that branch of commerce. Later on, about the middle of the eighteenth century, the droving of lean cattle into England was a means of profit. Meantime the system of cultivation was ot the rudest. A few acres round the steading, called the infield, were cropped year after year with little manuring, while the area beyond, called the outfield, was only cropped occasionally. There was no drainage and enclosures were unknown. Improvement came from the south. Sir Archibald Grant ot Monymusk and Mr Alexander Udny of Udny were pioneers of better things; they brought labourers and overseers from the Lothians and the south of England, to educate the people in new methods of culture. At first a landlord's, it by and by became a farmer's battle; and ultimately in the nineteenth century it was the farmers who did the reclaiming. But the landlords set a good example by sowing grasses and turnips.

Near Aberdeen, a boulder-strewn wilderness was con-

verted into fertile fields. The town feued the lands and
the feuars cleared away the stones, which they sold and
shipped to London for paving purposes; the process of
clearing cost as much as £100 an acre, a fourth of this
being recovered by the sale of the stones. This is typical
of what was done elsewhere. Gradually the bleak moors
were absorbed. A famine in 1782 opened the eyes of all
concerned. Hitherto there was not as much as 200 acres
in turnips. Hitherto also the heavy work-oxen, ten or
twelve of them dragging a primitive and shallow plough,
at a slow pace and in a serpentine furrow, had been
imported from the south. Now they began to be bred
on the spot. By and by cattle grew in numbers; by
and by, two horses superseded the team of oxen in the
plough.

But the chief factor in evolving Aberdeenshire into a
cattle-rearing and beef-producing county was the turnip.
Till turnips began to be grown in a large acreage, no
provision was possible for the cattle in winter. Hence
the beasts had to be disposed of in autumn. In 1820, as
many as 12,000 animals were sent in droves to England.
The advent of steam navigation in 1827 ended the
droving. Then began the trade in fat cattle, but it was
years before the county gained its laurels as the chief
purveyor of "prime Scots" and the roast beef of Old
England. The turnip held the key of the position; but
turnips will not grow well without manure. The canal
between Aberdeen and Inverurie carried great quantities
of crushed bones and guano to raise this important
crop.

Cattle-breeding began with McCombie of Tillyfour and the Cruickshanks of Sittyton, one with the native black-polled cattle—the Aberdeen-Angus—and the others with shorthorns. By dint of careful selection, great progress was made in improving not only the symmetry of the beasts but their size and beefy qualities. There

Aberdeen-Angus Bull

began a furore for cattle-rearing and prizes taken at Smithfield made Aberdeen famous. Railway transit came in as an additional help, and to-day the Christmas market never fails to give its top prices for Aberdeenshire beef.

Every year the beef of 60,000 cattle leaves the county for the southern markets, chiefly London ; this in addition to supplying local needs, and Aberdeen has now 162,000 of a population. Cattle-rearing and cattle-feeding are therefore at the backbone of Aberdeenshire agriculture.

A recent development is the export of pure-bred short-horns to America, more especially the Argentine Republic, for breeding purposes. As much as £1000 has been given for a young bull, in this connection.

Aberdeen Shorthorn Bull

In the matter of fruit culture, Aberdeen is far behind Perthshire and Lanark, which have a richer soil and a superior climate. But the Aberdeen strawberries, grown mostly on Deeside, are noted for size and flavour. In 1909 only 219 acres were devoted to this crop. The cultivation of raspberries, which is so great a feature of lower Perthshire, has made only a beginning in Aberdeen, and the small profits that have come to southern growers of this crop in recent years have acted as a deterrent, in its extension.

12. The Granite Industry.

Aberdeen has long been known as "The Granite City." It is built of granite, chiefly from its great quarries at Rubislaw. The granite is a light grey, somewhat different in texture and grain from another grey granite much in vogue, that of Kemnay on Donside. There are many quarries in the county, and each has its distinctive colouring. The Peterhead stone is red; Corrennie is also red but of a lighter hue. The granite industry has made great strides of recent years. The modern appliances for boring the rock by steam drills, the use of dynamite and other explosives for blasting, as well as the devices for hoisting and conveying stones from the well of the quarry to the upper levels by means of Blondins have all revolutionised the art of quarrying.

It was long before Aberdeen people realised the value of the local rocks for building purposes. The stone used in the early ecclesiastical buildings was sandstone, which was imported by sea from Morayshire and the Firth of Forth. The beginnings of St Machar Cathedral and the old church of St Nicholas as well as the church of Greyfriars, built early in the sixteenth century and recently demolished, were all of sandstone. Not till the seventeenth

Granite Quarry, Kemnay

century was granite utilised.   At first the surface stones
were taken, then quarrying began about 1604, but little
was done till 1725.   Between 1780 and 1790 as many as
600 men were employed in the Aberdeen quarries.   Great
engineering works such as the Bell Rock Lighthouse,
the Thames Embankment, the foundations of Waterloo
Bridge, the Forth Bridge and London Bridge, where
great durability and solidity are necessary, were made
possible by the use of huge blocks from Aberdeenshire.
The polishing of the stone made a beginning in 1820, and
now a great export trade in polished work for staircases,
house fronts, facades, fountains and other ornamental
purposes is carried on between the county and America
as well as the British Colonies.

Apart from building purposes, granite slabs are largely
used for headstones in graveyards.   This monumental
department employs a great number of skilled workmen.
There are over 80 granite-polishing yards in Aberdeen.
Here too the modern methods of cutting and polishing
the stones by machinery and pneumatic tools have greatly
reduced the manual labour as well as improved the
character of the work.   Unfortunately the export trade
in these monumental stones has somewhat declined owing
to prohibitive tariffs.   In 1896 America took £55,452
worth of finished stones ; in 1909 the value had fallen to
£38,000.   The tariffs in France have also been against
the trade, but an average of nearly 10,000 tons is sent to
continental countries.   Strangely enough, granite in the
raw state is itself imported to Aberdeen.   Swedish, Nor-
wegian and German granites are brought to Aberdeen, to

Granite Works, Aberdeen

be shaped and polished. These have a grain and colouring absolutely different from what is characteristic of the native stone, and the taste for novelty and variety has prompted their importation. In 1909 as much as 27,308 tons were imported in this way. Celtic and Runic crosses, recumbent tombs, and statuary are common as exports.

The stone is also used for the humbler purpose of street paving and is shipped to London and other ports in blocks of regular and recognised sizes. These are called "setts," and of them 30,000 tons are annually transmitted to the south. Stones of a larger size are also exported for use as pavement kerbs.

The presence of quarries is not so detrimental to the atmosphere and the landscape as coal mines, and yet the heaps of *débris*, of waste and useless stone piled up in great sloping ridges near the granite quarries, are undoubtedly an eyesore. To-day a means has been found whereby this blot on the landscape is partially removed. The waste *débris* is now crushed by special machinery into granite meal and gravel, and used as a surface dressing for walks and garden paths—a purpose it serves admirably, being both cleanly and easily dried. Not only so but great quantities of the waste are ground to fine powder, and after being mixed with cement and treated to great pressure become adamant blocks for pavements. These adamant blocks have now superseded the ordinary concrete pavement just as it superseded the use of solid granite blocks and Caithness flags. This ingenious utilisation of the waste has solved the problem which was beginning to face many of the larger quarries, namely, how they could dispose of their waste without burying valuable agricultural land under its mass.

Granite is the only mineral worthy of mention found in the county. Limestone exists in considerable quantities here and there, but as a rule it is too far from the railway routes to be profitably worked. It is, however, burned locally and applied to arable land as a manure. In the upper reaches of Strathdon, lime-kilns are numerous. By means of peat from the adjoining mosses the limestone was regularly burned half a century ago. To-day the practice is dwindling. A unique mineral deposit called Kieselguhr is found in considerable quantity in the peat-mosses of Dinnet, on Deeside. It is really the fossil remains of diatoms, and consists almost entirely of silica with a trace of lime and iron. When dried it is used as a polishing powder for steel, silver and other metals; but its chief use is in the manufacture of dynamite, of which it is the absorbent basis. It absorbs from three to four times its own weight of nitro-glycerine, which is the active property in dynamite. As found in the moss it is a layer two feet thick of cheesy light coloured matter, which is cut out into oblong pieces like peats. When these are dried, they become lighter in colour and ash-like in character. The Dinnet deposits are the only deposits of the kind in the country. Inferior beds are found in Skye. The industry employs 50 hands during the summer months, and has been in operation for 28 years. The beds show no sign of exhaustion as yet, and the demand for the substance is on the increase.

13. Other Industries. Paper, Wool, Combs.

The industries apart from agriculture, work in granite, and the fisheries are mostly concentrated in and around the chief city. These, although numerous, are not carried on in a large way, but they are varied; and there is this advantage in the eggs not being all in one basket that when depression attacks one trade, its effect is only partial and does not affect business as a whole. Paper, combs, wool, soap are all manufactured. The first of these engages four large establishments on the Don and one on the Dee at Culter. Writing paper and the paper used for the daily press and magazines as well as the coarser kinds of packing paper are all made in considerable quantity. Esparto grass and wood pulp are imported in connection with this industry. Comb-making is also carried on, and the factory in Aberdeen is the largest of its kind in the kingdom.

Textile fabrics are still produced, but the progress made in these is not to be compared with the advances made in the south of Scotland, where coal is cheap. Weaving was introduced at an early period by Flemish settlers, who made coarse linens and woollens till the end of the sixteenth century, when "grograms" and worsteds, broadcloth and friezes were added. Provost Alexander Jaffray the elder in 1636 established a house of correction —the prototype of the modern reformatory—where beggars and disorderly persons were employed in the manufacture of broadcloths, kerseys and other stuffs. A record of this novelty in discipline survives in the Aberdeen street called Correction Wynd.

In 1703 a joint-stock company was formed for woollen manufactures at Gordon's Mills on the Don, where a fulling-mill had existed for generations, and where the making of paper had been initiated a few years earlier. The Gordon's Mills developed the manufacture of cloths of a higher quality, half-silk serges, damask and plush, and skilled workmen were brought from France to guide and instruct the operatives. To-day high-class tweeds are made at Grandholm, and such is the reputation of these goods for quality and durability that in spite of high tariffs they make their way into America, where they command a large sale at prices more than double of the home prices.

In the olden days the cloth sold in the home markets was a product of domestic industry. The farmers' daughters spun the wool of their own sheep into yarn, which was sent to county weavers to be made into cloth. Aberdeenshire serge made in this way was sold at fairs and was hawked about the county by travelling packmen.

The hosiery trade was worked on similar lines. The wool was converted into worsted by rock and spindle, and the worsted was knitted into stockings by the women and girls of the rural population. One man employed as many as 400 knitters and spinners. In the latter half of the eighteenth century this industry brought from £100,000 to £120,000 into Aberdeenshire every year.

Stockings were made of such fineness that they cost 20*s.* a pair and occasional rarities were sold at four or five guineas. In 1771 twenty-two mercantile houses were engaged in the export trade of these goods, which went chiefly to Holland. The merchants attended the weekly markets and country fairs, where they purchased the products of the knitters' labour. Such work provided a source of income to the rural population and was indirectly the means of increasing the number of small holdings. These were multiplied of set purpose to keep the industrious element in the population within the county. The invention of the stocking-frame together with the dislocation of trade due to the Napoleonic wars made the trade unremunerative and it came to an end with the eighteenth century.

The linen trade began in 1737 at Huntly, where an Irishman under the patronage and encouragement of the Duke of Gordon manufactured yarn and exported it to England and the southern Scottish towns. Silk stockings were also made there. By and by linen works sprang up on the Don at Gordon's Mills and Grandholm as well as within Aberdeen itself. The linen trade in the form of spinning and hand-loom weaving was carried on in most of the towns and villages of the county and several new villages grew up in consequence, such as Cuminestown, New Byth, Strichen, New Pitsligo, Stuartfield and Fetterangus, in some of which the manufacture is continued on a small scale to this day. Much flax was grown in the county for a time to minister to this industry, but gradually the crop disappeared as fibre of better quality was imported from Holland. Yet the spinning of linen yarn was widely practised as a domestic industry when the woollen trade declined, and every farmer's daughter made a point of spinning her own linen as the nucleus of her future house-furnishings. The linen trade, except as regards coarser materials such as sacking, has decayed. There is still a jute factory in Aberdeen.

Another industry which employs a large number of hands is the preserving of meat, fish, fruit, and vegetables. There are several of these preserving works in Aberdeen. Dried and smoked haddocks, usually called "Finnan Haddies," from the village of Findon on the Kincardineshire coast, are one of the specialities of Aberdeen. They had at one time a great vogue, and are still largely in demand though the quality has fallen off by the adoption of a simpler and less expensive method of treatment.

Ship-building is another industry long established at Aberdeen. In the days before iron steamships, fleets of swift-sailing vessels known as "Aberdeen Clippers" were built on the Dee and made record voyages to China in the tea trade and to Australia. The industry of to-day is concerned, for the greater part, with the building of trawlers and other fishing craft, but occasionally an ocean going steamer is launched. The trade is meantime suffering from depression.

Other industries well represented are soap and candle making, coach and motor-car building, iron-founding and engineering, rope and twine making, the manufacture of

Making smoked haddocks, Aberdeen

chemicals, colours and aerated waters. Besides, Aberdeen is a great printing centre and many of the books issued by London publishers are printed by local presses.

14. Fisheries.

During the last quarter of a century the fishing industry has made great strides, the value of fish landed in Scotland having more than doubled in that period. Nowhere has the impetus been more felt than in Aberdeenshire, which now contributes as much as one-third of the whole product of Scotch fisheries. Since 1886 the weight of fish caught round the Scottish coast has increased from five million hundredweights to over nine millions, while the money value has risen in even a greater proportion from £1,403,391, to £3,149,127. To these totals Aberdeen alone has contributed over a hundred thousand tons of white fish (excluding herrings), valued at over a million pounds sterling. Peterhead and Fraserburgh are also contributors especially as regards herrings, the former landing 739,878 hundredweights and Fraserburgh very nearly a similar quantity. These three ports amongst them account for one half of all the fish landed at Scottish ports. When we consider the number of persons collaterally employed in handling this enormous quantity of merchandise, the coopers, cleaners, packers, basket makers, boat-builders, makers of nets, clerks and so on, apart altogether from the army of fishermen employed in catching the fish, we see how far-reaching this industry is, not merely in increasing the food-supply of the country but in providing profitable employment for the population. At Aberdeen, it is estimated, 13,512 persons are so employed and at the other two ports combined, almost a similar number.

There are two great branches of the fishing industry —herrings and white fish. The herrings are caught for the most part, though not exclusively, in the summer, July being the great month. They are captured with nets mostly by steam-drifters as they are called, but also to some extent by the ordinary sailing boats of a smaller size than the drifters. Fraserburgh and Peterhead land in each case double the weight of herrings that come to Aberdeen. In recent years a beginning has been made in May and June with gratifying success, but July and August give the maximum returns. Later on in the year, when the shoals have moved along the coast southwards, the herring fleets follow them thither, to North Shields and Hartlepool, to Yarmouth and Lowestoft; and bands of curers, coopers and workers migrate in hundreds from one port to another, employing themselves in curing the fish. The bulk of the herrings are cured by salting, and are then exported to Germany and Russia, where they are much in demand.

Even more important is the white fishing. Aberdeen is here pre-eminent, being perhaps the most important fishing centre in the world. The total catch for Scotland in 1909 was short of three million hundredweights, of which Aberdeen with its large fleet of trawlers and steam-liners accounted for 67 per cent. The most important of the so-called round fish is the haddock, of which over a million hundredweights are landed in Scotland, Aberdeen contributing the lion's share, three-fourths of the whole. Next to the haddock comes the cod, of which nearly three hundred and forty thousand hundredweights were handled in the Aberdeen fish market. The next fish is ling, and then come whitings, saithe, torsk, conger-eels. The flat-fish are also important, plaice, witches, megrims, halibut, lemon soles and turbot. This last is the scarcest and most highly prized of all flat fish, and commands a price next to that given for salmon. Ling and halibut are still mostly caught by hook and line; the turbot and the lemon sole on the other hand are distinctively the product of the trawl net and were little known until trawling was begun.

A certain small percentage of this great weight of fish is consumed locally, but the great bulk of it is packed in ice and dispatched by swift passenger trains to the southern markets. The Aberdeen fish market, extending for half a mile along the west and north sides of the Albert Basin (originally the bed of the Dee) is the property of the Town Corporation and is capable of dealing with large catches. As much as 760 tons of fish have been exposed on its concrete floor in a single day. In the early morning the place is one of the sights of the city, with the larger fish laid out in symmetrical rows on the pavement, and the smaller fish—haddocks, whiting and soles—in boxes arranged for the auction sale at 8 a.m.

The majority of the fishing craft are still sailing vessels, but steam-drifters and motor-boats and steam-

Fish Market, Aberdeen

trawlers are gradually driving the ordinary sail-boats from the trade, just as the trawl net is superseding the old-fashioned mode of fishing with set lines. Still about

Fishwives, The Green, Aberdeen

86 per cent. of the number of boats employed is made up of sailing vessels, but the tonnage is relatively small. The quantity of fish caught by hook and line is only one-tenth of the whole.

North Harbour, Peterhead

Herring boats at Fraserburgh

The amount of capital invested in boats and fishing gear for all Scotland is estimated by the Fishery Board at over £5,000,000. Of this total, Aberdeenshire claims very nearly two millions. It is now the case that the value of fish discharged at the fish market of Aberdeen is as great as the yearly value of the agricultural land of the whole county—truly a marvellous revolution.

Fishing Fleet going out, Aberdeen

The herring fishery was prosecuted off the Scottish coast by the Dutch, long before the Scotch could be induced to take part in it. Many futile attempts were made to exploit the industry but little came of them till the nineteenth century. A beginning was made at Peterhead in 1820 and at Fraserburgh a little earlier.

Aberdeen followed in 1836 but no great development took place till 1870. The first trawler came on the scene in 1882 ; to-day there are over 200 local vessels of this type besides many from other ports.

The salmon fishery has long been famous and at one time was relatively a source of much greater revenue than at present. It still yields a considerable annual surplus to the Corporation funds, but has been eclipsed by the growth of other fisheries. The rateable value of the salmon fishings on the Dee is nearly £19,000 ; those of the other salmon rivers—the Don, Ythan and Ugie— being much less. The fish are caught by fixed engines in the sea—stake-nets and bag-nets—set within a statutory radius of the river mouth, and by sweep- or drag-nets in the tidal reaches of the rivers. A good many fish are caught by rod and line throughout the whole course of the rivers but angling is not the commercial side of salmon-fishing.

15. Shipping and Trade.

Aberdeenshire has practically but three ports—Fraserburgh, Peterhead and Aberdeen. The herring fishing with its concomitant activities absorbs the energies of the two former so far as shipping is concerned, but Aberdeen having to serve a larger and wider area than these two northern burghs has developed a range of docks of considerable extent and importance. During the last forty years the Harbour Commissioners have spent £3,000,000 in improving the harbour, increasing the wharfage, adding break-waters, diverting the course of the Dee, deepening the entrance channel, forming a graving dock and so forth. Still, in spite of these outlays, Aberdeen, which has been a port for centuries, has hardly grown in shipping proportionately to its growth in other respects. The

At the docks, Aberdeen

reason is that, except fish, granite and agricultural products, the city has nothing of much moment to export.

Exclusive of fishing vessels the tonnage of home and foreign going vessels was in 1882, 587,173 ; in 1909 it had advanced to over a million, hardly doubling itself in 27 years. While its imports have gone up from 522,544 tons in 1882 to 1,165,060 in 1909, the exports have made only a very slight advance. The chief export is herrings, and last year nearly 100,000 tons of these, salted and packed in barrels, were sent by sea. The fresh fish are dispatched by rail. Stones in the form of granite, either polished for monumental purposes or in setts and kerbs for paving, account for 50,000 tons. The remainder (of 210,554 tons) is made up by oats, barley, oatmeal, paper, preserved provisions, whisky, manures, flax and cotton fabrics, woollen cloth, cattle and horses, butter and eggs, salmon and pine-wood.

The trade is mostly a coasting trade and more an import than an export one. Coal is the chief article of import, 600,000 tons being discharged in a year. Besides coal, esparto grass, wood-pulp and rags for paper-making, foreign granite in the rough state sent to be polished, flour, maize, linseed, the horns of cattle used for comb-making, and the salt used in fish-curing, are the chief materials landed on the Aberdeen quays. Aberdeen being the distributing centre for the county, and all the railway routes focussing in it, the coal and the building materials not produced in the district, such as lime, slate and cement, all pass this way, while the tea and sugar, the tobacco and other articles of daily use, also arrive mostly by the harbour.

There are regular lines of steamers between Aberdeen and the following ports : London, Newcastle, Hull, Liverpool, Glasgow and Leith, as also with continental towns such as Hamburg, Rotterdam and Christiania.

16. History of the County.

Standing remote from the centre of the country, Aberdeenshire has not been fated to figure largely in general history. The story of its own evolution from poverty to prosperity is an interesting one, but it is only now and again that the county is involved in the main current of the history of Scotland.

If the Romans ever visited it, which is highly doubtful, they left no convincing evidence of their stay. Of positive Roman influence no indication has survived, and no conquest of the district can have taken place. The only records of the early inhabitants of the district—usually called Picts—are the Eirde houses, the lake dwellings or crannogs, the hill forts or duns, the " Druidical " circles and standing stones and the flint arrow-heads, all of which will be dealt with in a later chapter.

Christianity had reached the south of Scotland before the Romans left early in the fifth century. The first missionary who crossed the Mounth was St Ternan, whose name survives at Banchory-Ternan on the Dee, the place of his death. St Kentigern or St Mungo, the patron saint of Glasgow, had a church dedicated to him at Glengairn. St Kentigern belonged to the sixth century, and was therefore a contemporary of St Columba, who christianised Aberdeenshire from Iona. In this way two great currents met in the north-east. Columba accompanied by his disciple Drostan first appeared at Aberdour on the northern coast. From Aberdour he passed on through Buchan, and having established the Monastery of Deer and left Drostan

in charge, moved on to other fields of labour.   His name
survives in the fishing village of St Combs.   He is the
tutelar saint of Belhelvie, and the churches of New Machar
and Daviot were dedicated to him.   These facts indicate
the mode in which Pictland was brought under the
influence of Christianity.

The next historical item worthy of mention is the
ravages of the Scandinavian Vikings.   The descents on
the coast of these sea-rovers were directed against the
monastic communities, which had gathered some wealth.
The Aberdeenshire coast, having few inlets convenient
for the entry of their long boats, was to a large extent
exempt from their raids, but in 1012 an expedition under
Cnut, son of Swegen, the king of Denmark, landed at
Cruden Bay.

Another fact of interest is the death of Macbeth, who
for seventeen years had by the help of Thorfinn, the
Scandinavian (whose name may be seen in the Deeside
town of Torphins), usurped the kingship of Scotland.
Malcolm Canmore led an army against him in 1057, and
gradually driving him north, beyond the Mounth, over-
took him at Lumphanan.   There Macbeth was slain.
A Macbeth's stone is said to mark the place where he
received his death-wound, and Macbeth's Cairn is marked
by a clump of trees in the midst of cultivated land.   The
farm called Cairnbethie retains the echo of his name.
Kincardine O'Neil, where Malcolm awaited the result of
the conflict, commands the ford of the Dee on the ancient
route of travel from south to north across the Cairn-o-
Mounth.

Malcolm shortly after passed through Aberdeenshire at the head of an expedition against the Celtic population which had supported Macbeth. The Norman Conquest, nine years thereafter, was the occasion of Anglo-Saxon settlements in the county. The court of Malcolm and Queen Margaret became a centre of Anglo-Saxon influence. The old Gaelic language gave way before the new Teutonic speech. The Celtic population made various attempts to recover the power that was fast slipping from their hands. Malcolm headed a second expedition to Aberdeenshire in 1078, and on that occasion granted the lands of Monymusk and Keig to the church of St Andrews. He is said to have had a hunting-seat in the forest of Mar, and the ruined castle of Kindrochit in the village of Braemar is associated with this fact.

The earliest mention of Aberdeen is in a charter of Alexander I, granting to the monks of Scone a dwelling in each of the principal towns—one of which is Aberdeen. A stream of Anglo-Saxons, Flemings and Scandinavians had been gradually flowing towards the settlement at the mouth of the Dee, where they pursued their handicrafts and established trade with other ports. William the Lion frequently visited the town and ultimately built a royal residence, which after a time was gifted to the Trinity or Red Friars for a monastery. The bishopric of Aberdeen dates from 1150.

Edward I of England in 1296 at the head of a large army paid these northern parts a visit. He entered the county by the road leading from Glenbervie to Durris, whence he proceeded to Aberdeen, exacting homage from

the burghers during his five days' stay.   From Aberdeen
he went to Kintore and Fyvie and on to Speyside, return-
ing by the Cabrach, Kildrummy, Kincardine O'Neil and
the Cairn-o-Mounth.

The next year Wallace, in his patriotic efforts to clear
the country from English domination, surprised Edward's
garrison at Aberdeen, but unable to effect anything, hastily
withdrew from the neighbourhood.   Edward was back
in Aberdeen in 1303 and paid another visit to Kildrummy
Castle, then in the possession of Bruce.   Then Bruce,
having fled from the English court and assassinated the
Red Comyn at Dumfries, was crowned at Scone and the
long struggle for national independence began in earnest.
In 1307 he came to Aberdeen, which was favourable to
his cause.   At Barra, not far from Inverurie and Old
Meldrum, his forces met those of the Earl of Buchan
(John Comyn) and defeated them (1308).   It was not a
great battle in itself, but its consequences were important.
It marked the turn of the tide in the national cause.
The Buchan district, in which the battle took place,
had long been identified with the powerful family of the
Comyns ; and after his victory at Barra, Bruce devastated
the district with relentless fury.   This " harrying of
Buchan," as it has been called, is referred to by Barbour
as an event bemoaned for more than fifty years.   The
family of the Comyns was crushed, and their influence,
which had been liberal and considerate to the native race
of Celts, came to an end.   The whole of the north-east
turned to Bruce's support, and in a short time all Edward's
garrisons disappeared.   This upheaval created a fresh

partition of the lands of Aberdeenshire. New families such as the Hays, the Frasers, the Gordons and the Irvines, were rewarded for faithful service by grants of land. The re-settlement of the county from non-Celtic sources accentuated the Teutonic element in the county. After Bannockburn, Bruce rewarded Aberdeen itself for its support by granting to the burgesses the burgh as well as the forest of Stocket.

The great event of the fifteenth century was the Battle of Harlaw, which took place in 1411 at no great distance from the site of the Battle of Barra. It was really a conflict between Celt and Saxon, and was a despairing effort on the part of the dispossessed native population to re-establish themselves in the Lowlands. The Highlanders were led by Donald of the Isles, who gathering the clansmen of the northern Hebrides, Ross and Lochaber, and sweeping through Moray and Strath-bogie, arrived at the Garioch on his way to Aberdeen. The burghers placed themselves under the leadership of the Earl of Mar (Alexander Stewart, son of the Wolf of Badenoch), a soldier who had seen much service in various parts of the world. The provost of the city, Robert Davidson, led forth a body of his fellow-citizens and joined Mar's forces at Inverurie, within three miles of the Highlanders' camp. The two forces were unequally matched—Donald having 10,000 men and Mar only a tenth of that number, but of these many were mail-clad knights on horseback and armed with spears. It was a fiercely contested battle and lasted till the darkness of a July night. The slaughter on both sides was great, but the tide of barbarism was driven back. The Highlanders retreated whence they came and the county of Aberdeen was saved from the imminent peril of a Celtic recrudescence. This is the only really memorable battle associated with Aberdeenshire soil. Its "red" field, on which so many prominent citizens shed their life-blood (Provost Davidson and Sir Alexander Irvine of Drum being of the number), was long remembered as a dreary and costly victory.

Another battle of much less significance was that of Corrichie, fought in Queen Mary's reign in 1562 on the eastern slope of the Hill of Fare, not far from Banchory. It was a contest between James Stewart (the Regent Murray, and half-brother of the Queen) and the Earl of Huntly. Huntly was defeated and slain, and his son, Sir John Gordon, who was taken prisoner, was afterwards executed at Aberdeen. Queen Mary, it is said, was a spectator both of the battle and of the execution.

In the seventeenth century, at the beginning of the Covenanting "troubles," Aberdeenshire gained a certain notoriety as being the place where the sword was first drawn. In 1639 the Covenanters mustered at Turriff under Montrose, to the number of 800. The Royalist party under the Earl of Huntly, to the number of 2000 but poorly armed, marched to the town with the intention of preventing the Covenanters from meeting, but they were already in possession, and when Huntly's party saw how matters stood, they passed on, the two forces surveying each other at close quarters without hostile act or word. This bloodless affair is known as the first

Raid of Turriff. A few weeks later a somewhat similar encounter took place, when the Covenanters, completely surprised, fled without striking a blow. The loss on either side was trifling, still some blood was actually shed, and the Trot of Turriff, as it was called, became the first incident in a long line of mighty events.

Montrose, both when he was leading on the side of the Covenant, and later when he became a Royalist leader, paid several visits to Aberdeen, which, although supporting the Royalist cause, suffered exactions from both parties. In 1644 Montrose made a forcible entry of the town, which resulted in the death of 150 Covenanters, and in the plundering of the city. Later on, after his victory over Argyll at Inverlochy, Montrose gained a success for the Royalist cause at Alford (1645).

In 1650, after the execution of Charles I, his son Charles II landed at Speymouth, and on his way south to be crowned at Scone, visited Aberdeen, where he was received with every manifestation of loyalty and good-will. The next year General Monk paid the town a visit, and left an English garrison, which remained till 1659. The Restoration was hailed with rejoicing : the Revolution with dislike. Yet at the Rebellion of 1715 scant enthusiasm was roused for the cause of the Pretender, who himself passed through the city on his way from Peterhead to Fetteresso. In the thirty years that passed before the second Jacobite Rebellion, public sentiment had grown more favourable to the reigning House. The '45 therefore received little support in Aberdeenshire. A few of the old county families threw in their lot with the Prince, but the general body of the people were averse to taking arms. The Duke of Cumberland, on his way north to meet Prince Charlie at Culloden, remained with his army six weeks in the city; when he started on his northward march through Old Meldrum, Turriff and Banff, he left a garrison of 200 men in Robert Gordon's Hospital, lately built but not yet opened. After Culloden small pickets of troops were stationed in the Highland districts of the county, to suppress the practice of cattle-lifting. Braemar Castle and Corgarff Castle in the upper reaches of the Dee and the Don still bear evidence of their use as garrison forts. The problem of dealing with the inhabitants of the higher glens, where agriculture was useless, and where the habits of the people prompted to raiding and to rebellion, was solved by enlisting the young men in the British Army. The Black Watch (42nd) as reorganised (1758) and a regiment of Gordon Highlanders (1759) were largely recruited from West Aberdeenshire, and this happy solution closed the military history of the district.

17. Antiquities—Circles, Sculptured Stones, Crannogs, Forts.

Aberdeenshire is particularly rich in stone-circles. No fewer than 175 of them have been recorded as existing in the district. Unfortunately many of them entirely disappeared when the sites were turned to agricultural uses; others have been mutilated, and owing to the removal of some of the stones, stand incomplete ; a few have been untouched, and from these we may judge what the others were like. One of the best preserved is that at Parkhouse, a mile south-west of the Abbey of Deer. A circle of great blocks of stone, irregular and of unequal height, some standing erect, some evidently fallen down, is the general feature. Sometimes inside the circle, but more usually in the circumference of the circle itself, there is one conspicuously larger stone, in a recumbent position. This it has been usual to call the rostrum or altar stone. It is well marked at Parkhouse, being 14

White Cow Wood Cairn Circle; View from the S.W.
From *Proceedings of the Society of Antiquaries of Scotland*, 1903-4

feet 9 inches long, 5 feet 9 inches high, and estimated to weigh 20 tons. The so-called rostrum is usually on the south side of the circle and the stones facing it on the north are of smaller size.

The size of the circles varies, the largest being over 60 feet in diameter, the smaller ones less than 30. Parkhouse measures 50 feet. They are found all over the county, in the valley of the Dee, in the valley of the Don at Alford, Inverurie and Dyce, as well as in Auchterless, Methlick, Crimond and Lonmay. The recumbent stone

Palaeolithic Flint Implement
(From Kent's Cavern, Torquay)

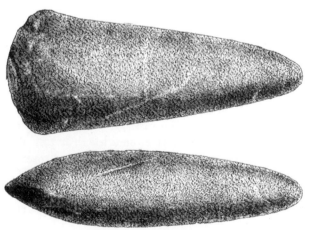

Neolithic Celt of Greenstone
(From Bridlington, Yorks.)

is invariably a feature of the larger circles. One of the largest is in the circle at Old Keig in Alford—a huge monolith computed to be 30 tons in weight. Other good examples are at Auchquorthies, Fetternear and at Balquhain near Inveramsay.

In the smaller and simpler circles, there is no recumbent stone, and the blocks are of more uniform height.

What the circles were used for is still a matter of dispute. They have for long been called "Druidical" circles, and the received opinion was that they were places of worship, the recumbent stone being the altar. But there is no certitude in this view; and, indeed, the fact that several exist at no great distance from each other (more than a dozen are located in Deer) would seem to be adverse to it. They were certainly used as places of burying, and some antiquarians hold that they were the burying grounds of the people of the Bronze Age. A later theory is that they were intended to be astronomical clocks to a people who knew nothing of the length of the year, and who had no almanacs to guide them in the matter of the seasons. The stone-circles, however, still remain an unsolved problem.

Besides the circles, Aberdeenshire has another class of archaeological remains, called sculptured stones. These are of three kinds: (1) those with incised symbols only, (2) those with in addition Celtic ornament carved in relief, and (3) monuments with Celtic ornament in relief and no symbols. The first class is the only one largely represented in Aberdeenshire and a good many representatives are in existence. The symbols most com-

monly seen are the crescent and sceptre, the spectacles,
the mirror and comb, and the so-called "elephant" symbol,

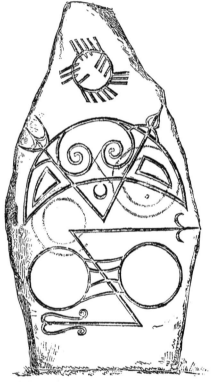

Stone at Logie, in the Garioch (4 feet high)
From Anderson's *Scot. in Early Ch. Times*, 2nd Series

a representation of a beast with long jaws, a crest and
scroll feet. Another is the serpent symbol. What the

symbols signify is still a mystery, but the fact that the stones with symbolism are unusually common in what was known as Northern Pictland seems to point to their being indigenous to that area. Out of 124 stones in the first class Aberdeenshire has 42. It would seem as if the county had been the focus where the symbolism originated. The richness of the locality round Kintore and Inverurie in symbol stones is taken to indicate that region as the centre from which they radiated.

Another form of archaeological remains found in the county is the Eirde or Earth-Houses. These are subterranean dwellings dug out of the ground and walled with unhewn, unmortared stones, each stone overlapping the one below until they meet at the top which is crowned with a larger flag-stone, or sometimes with wood. The probability is that in conjunction with the underground chambers there were huts above ground, which, being composed of wood, have now entirely disappeared. At many points in these earth-houses traces of fire and charcoal are to be seen, stones blackened by fire and layers of black ashes. In one at Loch Kinnord a piece of the upper stone of a quern as well as an angular piece of iron was found. It may be inferred that the inhabitants, whoever they were, were agriculturists, and that the period of occupation lasted down to the Iron Age. Specimens of these houses, which usually go by the local name of Picts' houses, are found in the neighbourhood of Loch Kinnord on Deeside, at Castle Newe on the Don, and at Parkhouse, not far from the circle already referred to.

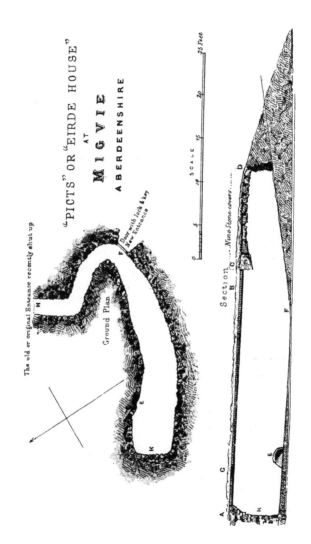

"PICTS" OR "EIRDE HOUSE"

AT

MIGVIE

ABERDEENSHIRE

The old or original Entrance recently shut up

Ground Plan

Door with lock & key

New Entrance

SCALE

Section ···· Nine Stone covers ·····

From *Proceedings of the Society of Antiquaries of Scotland*, Vol. v. 1865

The common notion of the purpose of these underground dwellings was that they were meant for hiding-places in which the inhabitants took refuge when unable to resist their enemies in the open, but if, as has now been discovered, they were associated with wooden erections above ground, they could not have served this purpose. On the surface beside them were other houses, cattlefolds and other enclosures; once an enemy was in possession of these, he could hardly miss the earth-houses. Moreover, the inhabitants, if discovered, were in a trap from which there was no escape. It is more probable that the dwellings were adjuncts of some unknown kind to the huts on the surface. The fact that pottery and bronze armlets have been unearthed from these underground caverns proves that the earth-dwellers had reached a certain advancement in civilisation. They reared domestic animals, wove cloth and sewed it, and manufactured pottery. They used iron for cutting weapons and bronze for ornament, and must have possesssed a wonderfully high standard of taste and manual skill.

Along with the earth-houses at Kinnord are found crannogs or lake-dwellings. Artificial islands were created in the loch by forming a raft of logs, upon which layers of stones and other logs were deposited. As fresh materials were added the raft gradually sank till it rested on the bottom. The sides were afterwards strengthened with the addition of stones and beams. In this way was formed what is called the Prison Island on Loch Kinnord. In all probability the other island in the same loch, the

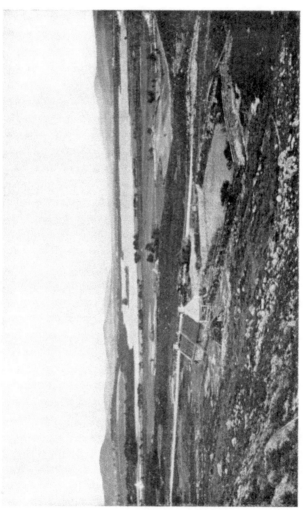

Loch Kinnord

Castle Island, may also be artificial, although it has usually been regarded as natural. Crannogs in pairs—one large and the other small—occur in several lochs.

A number of hill-forts, more or less disintegrated, are traceable in the higher ground in the vicinity of Lochs Kinnord and Davan. These show concentric lines of circumvallation, with stronger fortifications at various points. Vitrified forts, where the stones have been run together by the application of heat, are found at Dunnideer near Insch, and on the conical summit of Tap o' Noth near Rhynie. The Barmekin at Dunecht encloses an area of more than two acres, and consists of five concentric walls, three of earth-works and two of stone.

Numerous cairns, barrows or tumuli exist all over the county, at Aberdour on the coast, at Birse, Bourtie, Rhynie, Turriff, and elsewhere. Human remains have been found in most of these; and as a rule flint arrow-heads and other implements are also associated with them.

18. Architecture—(a) Ecclesiastical.

The history of Scotland from an architectural point of view does not reach very far back into the past. Till the tide of civilisation flowed into Scotland from the south in the eleventh and twelfth centuries, there existed in the country no architecture worthy of the name. When the Normans became the ruling power in Britain, they brought architectural ideas with them and these superseded the

crude attempts at church building hitherto made. The
Scottish churches built under the influence of Columba
were simple and rude, consisting of a small oblong chamber with a single door and a single window. The Norman
style, which obliterated these structures, dates from the
twelfth century and, being carried along the coast of lowland Scotland, gradually changed the manner of building.
It is characterised by simple, massive forms and especially
by arches of a semi-circular shape, sometimes enriched by
zig-zag, and by the use of nook shafts and cushion capitals.
Of this period the remains in Scotland are not numerous,
and they are very few in Aberdeenshire. The earliest
specimen we can point to is the ancient church of
Monymusk, which contains some Norman building incorporated in the modern church erected on the old site.
Monymusk is on Donside seven miles up the river from
Kintore. It is a place of great antiquity. The Culdees
first appear there in the twelfth century, and the Earl of
Mar built a convent for them on condition that they
should submit to canonical rule. The lower part of the
church tower and the chancel arch are of the Norman
style. The tower has been entirely rebuilt except the
lower doorway, which has a round arch-head with a hood
mould enclosing it. These small fragments suggest that
they were part of the convent erected by the Earl of Mar
very early in the thirteenth century.

The rounded arch gave place in the thirteenth century to the early Gothic, of which the most striking feature
is the pointed arch. This is the *First Pointed Period*.
Ornament was more general, the mouldings were richer

and more graceful and the foliage of trees was occasionally copied. The windows were narrow, lofty and pointed, giving an impression of space and lightness. Aberdeen-shire is too far north to have developed many examples of this early style, but it has some. The Abbey of Deer is perhaps the most ancient ecclesiastical building, but it is now a complete ruin, all the best parts of it having dis-appeared within the last fifty years. It was founded in the thirteenth century. Deer had been an ecclesiastical centre long before that time. The story goes that Columba and his pupil Drostan travelled from Iona to Aberdeenshire when Bede was Mormaer (Earl) of Buchan. They were first at Aberdour on the coast, but ultimately journeyed to Deer, where Columba requested the Mormaer to grant him a site for a church. At first the Mormaer refused, but his son fell ill and in consideration of the efficacy of the prayers of the two holy men in bringing the youth back to health, the Mormaer granted them the lands of Deer and this was probably the first place in Aberdeen-shire where a regular Christian church was erected. No trace of that church, built in the sixth century, is left.

The Abbey was an entirely different structure and not begun till early in the thirteenth century. It was founded by William Comyn, Earl of Buchan, and was really a Cistercian Abbey, originally occupied by monks sent from Kinloss. From the ruins now within the grounds of Pitfour House, it can be made out that the length of the building (nave and chancel) was 150 feet. A few mould-ings and the arches of some windows indicate that it belonged to the first pointed period. The building was

of red sandstone probably brought from New Byth, some
12 miles distant.   After the Reformation the Abbey fell
to decay and its walls became, as in many other cases, a
quarry from which other buildings were erected.   In 1809
the ruins were enclosed with a wall by the then proprietor,
Mr James Ferguson of Pitfour, but since then they have
dwindled.

No mention of Deer is possible without reference to
the famous *Book of Deer*—a manuscript volume of the
highest value, emanating not from the Abbey but from
Columba's monastery in the same region.   The book
was brought to light in 1860 by the late Mr Henry
Bradshaw, University Librarian at Cambridge.   It had
lain unrecognised in the Library since 1715.   It contains
the Gospel of St John and other portions of scripture in
the writing of the ninth century; but of even greater
importance is the fact that on its margins it contains
memoranda of grants to the monastery, made by Celtic
chiefs of Buchan and all written in Gaelic.   These
jottings are of the highest historical value.

Some traces of the Early Pointed style are found
in St Machar Cathedral (the greater part of which,
however, is much later).   The old church of Auchindoir
close to Craig Castle has a good doorway and other features
of this period.

From the middle of the fourteenth to the middle
of the fifteenth century (1350–1450) is in Scotland the
Middle Pointed Period.   The windows were made larger,
the vaulting and buttresses less heavy.   The Cathedral of
St Machar belongs in part to this time.   The legend goes

From *The Book of Deer*

that St Machar in obedience to the commands of Columba,
of whom he was a disciple, journeyed to Scotland and
at Old Aberdeen founded a church. This church in
the twelfth century became the seat of a bishopric
founded by David I. The original church was superseded probably about 1165, the only relic of this Norman
period being part of the abacus of a square pier. All
other traces of earlier work have vanished. In the fourteenth century Bishop Alexander Kyninmonth II rebuilt
the nave, partly of red sandstone with foliated capitals
of great beauty and decorated with naturalistic imitation
of leafage, one capital representing curly kail (colewort).
The same kind of decoration is seen in Melrose Abbey.
Later on the two impressive western towers, which are
to-day conspicuous objects in the eastern landscape to all
travellers northward-bound from Aberdeen, were added.
They form a granite mass of solid and substantial masonry,
and, being finished with machicolation, parapet-paths and
capehouses, were really like a castle in Early English
architecture. Still later on, in the sixteenth century,
Bishop Elphinstone, who founded the University of
Aberdeen, who built the first Bridge of Dee, and gave
a new choir to St Nicholas Church, completed the central
tower and placed in it fourteen bells "tuneable and
costly." The sandstone spires over the western towers
were added by Bishop Dunbar early in the sixteenth
century, in place of the original capehouses. The central
tower fell in 1688, crushing the transepts.

In 1560 the government ordered the destruction of
the altars, images and other monuments of the old faith,

St Machar Cathedral, Old Aberdeen

and this cathedral suffered with the rest. It was despoiled of all its costly ornaments and the choir was demolished. The roof was stripped of its lead and the bells were carried off. All that remains to-day is the nave (now the parish church), a south porch, the western towers and fragments of the transept walls, which contain tombs of

St Machar Cathedral (interior)

Bishop Lichtoun, Bishop Dunbar, and others. This is the only granite cathedral in the country, and, though dating from the Middle and late Pointed periods, has reminiscences of the Norman style in its short, massive cylindrical pillars and plain unadorned clerestory windows. Another feature is the great western window divided by six long shafts of stone. The finely carved pulpit now in the Chapel of King's College is a relic of the wood-carvings destroyed in 1649. The whole is extremely plain but highly impressive and imposing. Its flat panelled oak ceiling decorated with heraldic shields of various European kings, Pope Leo X, and Scottish ecclesiastics

King's College, Aberdeen University

and nobility (48 in all) is worthy of mention. This heraldic ceiling was restored in 1868–71.

Of later date is King's College Chapel, at no great distance from the Old Cathedral. It is a long, narrow but handsome building begun in 1500, shortly after the foundation of the University by Bishop Elphinstone. The chapel and its graceful tower are the oldest parts of the College buildings which had originally three towers.

The surviving one is a massive structure buttressed nearly to the top and bearing aloft a lantern of crossed rib arches, surmounted by a beautiful imperial crown with finial cross, somewhat resembling St Giles's in Edinburgh. The difference is that King's College has four ribs while

East and West Churches, Aberdeen

St Giles's has eight. The whole is of freestone from Morayshire. The entire building is a mixture of Scottish and French Gothic styles, and retains in the large western window the semi-circular arch, a peculiarity of Scottish Gothic throughout all periods. The canopied stalls and the screen of richly carved oak, Gothic in design and most beautifully handled, take a place among the finest pieces of mediaeval carved work existing in the British Empire. Their beauty and delicacy, according to Hill Burton, surpass all remains of a similar kind in Scotland. The chapel contains the tomb of Bishop Elphinstone. It was once highly ornamented, but meantime is covered with a plain marble slab. Its restoration is in prospect.

St Nicholas Church, Aberdeen (now the East and West Churches) contains in its transepts and groined crypt and in its wood-carving, interesting relics of twelfth, fifteenth and sixteenth century work. The nave was rebuilt in the Renaissance style of the time (1755).

Greyfriars Church, removed a few years ago to make way for the new front of Marischal College, was a pre-Reformation church, built by Alexander Galloway, Rector of Kinkell, early in the sixteenth century. Its chief features were its range of buttresses and a fine seven-light, traceried window.

The Protestant churches that succeeded these ancient buildings were inferior as architecture. It was only in the nineteenth century that taste began to revive and some attempt at grace and embellishment was made. Architects began to study old styles, and this combined with the increasing wealth of the country created a new standard in ecclesiastical requirements. To-day our churches tend to grow in architectural beauty.

19. Architecture—(b) Castellated.

The earliest fortifications in Scotland were earthen mounds, surrounded with wooden palisades. They were succeeded by stone and lime "keeps" built in imitation of Norman structures. The presence of the Normans in England during the eleventh and twelfth centuries drove the Saxon nobility northwards, and they were followed in turn by other Normans, who obtained possession of great tracts of country. The rectangular keeps of the Normans have in consequence formed the models on which most of the Scottish castles were constructed. In the thirteenth century there were castles at Strathbogie, Fyvie, Inverurie and Kildrummy. These have mostly been rebuilt in recent times and the more ancient parts have disappeared. The general idea in them all was a fortified enclosure usually quadrilateral. The walls of the enclosure were 7 to 9 feet thick and 20 to 30 feet high. The angles had round or square towers, and the walls had parapets and embrasures for defence and a continuous path round the top of the ramparts. The entrance was a wide gate guarded by a portcullis. The comparatively large area within the walls was intended to harbour the population of a district and to give temporary protection to their flocks and possessions in times of danger. Some of the finer examples, such as Kildrummy, closely resemble the splendid military buildings of France in the thirteenth century. One of the towers is usually larger than the others and forms the donjon or place of strength, to which retreat could be made as a last resort, when, during a siege, the enemy had gained a footing within the walls.

Kildrummy Castle

Kildrummy Castle is one of the finest and largest in Scotland, and even in its present ruinous condition gives an impression of grandeur and extent such as no other castle in Aberdeenshire can rival. It was built in the reign of Alexander II by Gilbert de Moravia, Bishop of Caithness. Situated near the river Don, some ten miles inland from Alford, and occupying a strong position on the top of a bank which slopes steeply to a burn on two sides, and protected on the other sides by an artificial fosse, it was a place of great strength. Its plan is an irregular quadrangle, the south front bulging out in the centre towards the gateway. It had six round towers, one at each angle and two at the gate. One of the corner towers—the Snow Tower—55 feet in diameter, was the donjon and contained the draw-well. The castle possessed a large courtyard, a great hall, and a chapel, of which the window of three tall lancets survives. It was built in the thirteenth century, and therefore belongs to the First Pointed Period. The stone used is a sandstone, probably taken from the quarries in the locality, where instead of the prevailing granite of Aberdeenshire a great band of sandstone occurs.

This famous castle passed through various vicissitudes. It was besieged in 1306 by Edward I of England and was gallantly defended, but, in consequence of a great conflagration, Nigel Bruce, King Robert's brother, who was acting as governor, yielded it to the English, he himself being made prisoner and ultimately executed. Some of the buildings date from this period, when it was rebuilt by the English, but it soon fell into Bruce's hands again. Twenty years after Bannockburn it was conferred on the Earl of Mar. The rebellion of 1715 was hatched within its walls. Thereafter being forfeited by Mar, it eventually came into the hands of the Gordons of Wardhouse. Recently it was purchased by Colonel Ogston, who has built a modern mansion-house close by and crossed the ravine with a bridge, an exact replica of the historic Bridge of Balgownie near Donmouth.

During the fourteenth century, Scotland, exhausted with the struggle for national independence, was unable to engage in extensive building. Beside, Bruce's policy was opposed to castle building, as such edifices were liable to be captured by the enemy and a secure footing thereby obtained. His policy was rather to strip the country, and to destroy everything in front of an invading army, with a view to starving it out. The houses of the peasantry were made of wood and could easily be restored when destroyed. The houses of the nobility took the form of square towers on the Norman model and all castles of the fourteenth century were on this simple plan—a square or oblong tower with very thick walls and defended from a parapeted path round the top of the tower. The angles were rounded or projected on corbels in the form of round bartizans. At first these parapets were open and machicolated. As time went on, the simple keep was extended by adding on a small wing at one corner, which gave the ground plan of the whole building the shape of the letter L. The entrance was then placed as a rule at the re-entering angle. Such keeps are usually spoken of as built on the L plan. The ground floor was vaulted and used for stores or stables and as accommodation for servants. The only communication between this and the first floor was a hatch. In early castles the principal entrance was often on the floor above the ground floor and was reached by a stair easily removed in time of danger. Access from one storey to another was by a corkscrew or newel stair at one corner in the thick wall. Thus constructed a tower could resist siege and fire, and even if taken, could not be easily damaged.

Of this kind of keep Aberdeenshire has many excellent examples, the most perfect, perhaps, being the Tower of Drum. It stands on a ridge overlooking the valley of the Dee. To the ancient keep built probably late in the thirteenth century was added a mansion-house on a different plan in 1619. The estate was granted to William de Irvine by Bruce in recognition of faithful service as secretary and armour-bearer. Previous to that, Drum was a royal forest and a hunting-seat of the king. The keep, which stands as solid and square to-day as it did six hundred years ago, is quadrilateral and the angles are rounded off. The entrance was at the level of the first floor. The main stair is a newel. In the lowest storey the walls are twelve feet thick, pierced with two narrow loops for light. In a recess is the well. On the top of the tower are battlements, the parapet resting on a corbel-table continued right round the building.

Hallforest near Kintore is an example of a fourteenth century keep. It was built by Bruce as a hunting-seat and bestowed on Sir Robert de Keith, the Marischal. It still belongs to the Kintore family but is now a ruin.

The fifteenth century brought a change in castle-building. The accommodation of the keeps was circumscribed and the paucity of rooms made privacy impossible.

One way of extending the space was, as we have said, by adding a wing at one corner. Another mode was to utilise the surrounding wall, for the keeps were generally guarded by a wall, which formed a courtyard or barmekin for stabling and offices. This was often of considerable extent and defended by towers. As the country progressed and manners improved, buildings were extended round the inside of the courtyard walls. In the sixteenth century the change went further and developed into the mansion-house built round a quadrangle. The building was first in the centre of the surrounding wall; ultimately the courtyard was absorbed and became the centre of the castle.

Balquhain Castle in Chapel of Garioch, two miles from Inverurie, was originally a keep like Drum, but being destroyed in 1526, it was rebuilt. Very little of it now remains but its massive, weather-stained walls have a commanding effect. The barmekin is still traceable. Queen Mary is said to have passed the night prior to the Battle of Corrichie at Balquhain. It was burned in 1746 by the Duke of Cumberland.

Many other castles on the same general plan are dotted up and down the county. Some are in ruins, some have been altered and added to on other lines, but the original keep is still a marked feature in most of them. Cairnbulg—recently restored—on the north-east coast has a keep of the fifteenth century with additions of a century later. Gight, now ruinous, but formerly celebrated for its great strength, occupies a fine site on the summit of the Braes of Gight, which rise abruptly from the bed of the Ythan. It also is a fifteenth century edifice built on the L plan. It has a historical interest as having once belonged to Lord Byron's mother, from whom it was purchased by the Earl of Aberdeen. Another of the same kind is Craig Castle in Auchindoir. It was completed in 1518 and is also on the L plan. So too is Fedderat in New Deer.

The Old House of Gight

In the sixteenth century the troubled reign of Queen Mary was unfavourable to architecture, but towards the end of it the rise of Renaissance art began to exert a decided influence, especially on details and internal furnishings, and in the next century gradually but completely dominated the spirit of the art. Another influence at work was the progress made in artillery. The ordinary castles

could not now resist artillery fire, and all attempts at
making them impregnable fortresses were abandoned, and
the only fortifications retained were such as would make
the buildings safe from sudden attack.   In consequence,
what had before been grim fortresses were now transformed into country mansions, whether on the keep or on
the quadrangle plan ; and sites were chosen as providing
shelter from the elements rather than defence against
human foes.   The Reformation, too, which secularised
the church lands and gave the lion's share to the nobility,
was a notable influence in revolutionising architecture.
The nobility being now more wealthy were enabled either
to extend their old mansions or to build new ones. Hence
the great development that took place in the quiet reign
of James VI.   The effect of the Union in 1603, which
drew many of the nobility to England, was civilising and
educative, and raised their ideas of house accommodation
as well as their standard of comfort and domestic amenity.

The change was of course gradual.   The old keeps
and the castles built round a courtyard were still in
evidence, but picturesque turrets corbelled out at every
angle of the building, slated, and terminating in fanciful
finials, became the rule. The lower walls were kept
plain, the ornamentation being lavishly crowded only on
the upper parts.   The roofs became high-pitched with
picturesque chimneys, dormer windows and crow-stepped
gables.   All these features so characteristic of the mansion-
houses of the fourth period (1542–1700) are well marked
in Craigievar, which is one of the best preserved castles of
the time.   Its ground plan is of the L type, but the turrets

and gables are corbelled out with ornamental mouldings and the upper part of the castle displays that profusion of

Craigievar Castle, Donside

sky-pointing pinnacles and multifarious parapets which mark the period. The same is seen at Crathes and at

Castle Fraser. The last is altogether an excellent specimen. It consists of a central oblong building with two towers at the diagonally opposite ends, one square and the other round, and is therefore a development into what has been called the Z plan or stepped plan—induced by the general

Crathes Castle, Kincardineshire

use of firearms in defence. Here, as at Craigievar, gargoyles originally used to carry off rain water from the roof are brought in as a piece of fanciful decoration, apart from any utilitarian purpose, and project from the walls at places where rain-spouts are irrelevant.

The castle has a secret chamber or "lug," in which

the master of the house could over-hear the conversation of his guests in the dining-hall. Nothing could better illustrate the treachery and cunning which had been bred by the difficulty of the times. Mr Skene, the friend of Sir Walter Scott, minutely investigated this contrivance as it

Castle Fraser

exists at Castle Fraser, and no doubt his account of this ingenious but dishonourable device for gaining illicit information suggested King James's "Lug," so happily described in *The Fortunes of Nigel*.

Castellated buildings of this class are so numerous in Aberdeenshire that it is possible to name only a few.

One of the finest is Fyvie Castle on the banks of the upper reaches of the Ythan in the very centre of the county. It is not like many others a ruin, but a mansion-house modernised in many respects, but still retaining all the picturesque features of the olden time. It occupies two sides of a quadrangle, with the principal front towards the south, one side being 147, the other 137 feet in frontage. At the three corners are massive square towers, with angle turrets and crow-stepped gables. Besides these towers, there are in the centre of the south front two other projecting towers, which at 42 feet from the ground are bridged by a connecting arch, eleven feet wide, the whole forming a grand and most impressive mass of masonry called the "Seton" tower, a magnificent centre to what is perhaps the most imposing front of any domestic edifice in Scotland. At the south-east corner is the "Preston" tower, built by Sir Henry Preston, and the earliest part of the building, dating from the four-teenth century. In the south-west stands the "Meldrum" tower, so-called from the succeeding proprietors (1440–1596). They erected this part and the whole range of the south front except the "Seton" tower already referred to, which is a later addition. The Setons succeeded the Meldrums and it is to Alexander Seton, Lord Fyvie and Earl of Dunfermline, that the castle owes its greatest splendours. Besides planning this tower, he ornamented the others with their turreted and ornate details. He also built the great stair-case, which is a triumph of architectural skill. It is a wheel or newel staircase of grand proportions, skilfully planned and as skilfully executed. The Gordon tower on the west was not added till the eighteenth century, by William, second son of the second Earl of Aberdeen. Its erection necessitated the destruction of the chapel. Here one may see how the

Fyvie Castle, South Front

Renaissance ideas were creeping in, especially the desire for balance and symmetry. Two of everything was beginning to be the rule. One wing must have another to balance with it ; one tower another to make a pair.

20. Architecture—(c) **Municipal.**

After a period of declining taste in architecture, a revival began early in the nineteenth century under the guidance of architects of genius such as Archibald Simpson and John and William Smith. A great improvement was thereby effected in the general aspect of the city of Aberdeen, and their good work has been enhanced by that of their successors. It is necessary to repeat that it was long before the local granite came to its own. The earlier buildings of importance were all of sandstone; to-day he would be a bold architect who suggested a sandstone building in Aberdeen. The use of granite exercises an indirect effect on architectural design. It lends itself to broad, classic, monumental and dignified effects, while its stubborn quality is a check against over-exuberance of detail, and fanciful, gimcrack trivialities. The plainness of the buildings was often remarked upon by strangers twenty years ago. The newer buildings are not without adornment.

The County and Municipal Buildings (or the Town-House as it is familiarly called) on the south side of Castle Street were opened in 1870. They form a magnificent pile which takes a high place amongst provincial town-halls, as regards both vigour and originality of treatment. The line of elliptical arches on the ground floor and of small arcaded windows in the floor above make an imposing front. The great tower, which rises to a height of 200 feet and dominates the whole city, has the castellated turrets which we have seen to be characteristic of Scottish architecture. It is curious to see how latter-day architects have not been able to get away from this feature. It is conspicuous even in such buildings as the Grammar School and the new Post Office. The

Municipal Buildings, Aberdeen, and Town Cross

Municipal tower, if somewhat heavy-looking, is on the whole effective. The small tower and spire on the east is the old Tolbooth tower, of the seventeenth century, preserved by being incorporated in the modern building.

The next public building that should be mentioned is Marischal College, recently enlarged at a cost of nearly

£250,000. This is undoubtedly the finest piece of modern architecture in the north of Scotland, and one of the most handsome and graceful in the kingdom. The College at the end of the nineteenth century was a work of the Gothic revival occupying three sides of a quadrangle, with a tower in the centre of one side. This tower has

Marischal College, Aberdeen

been remodelled and greatly heightened so that it is now a rival to the Municipal tower in the same street. It is known as the Mitchell tower, in compliment to the donor, the late Mr Charles Mitchell of Newcastle, whose name is also associated with the public or graduation Hall of the University. The old frontage of Marischal College was a desultory line of commonplace houses, through which by a narrow gateway entrance was gained to the quadrangle. These have all been cleared away and now a stately pile bristling with ornate pinnacles that sparkle in the sun fills the whole length of 400 feet.

No less impressive than the delicately chiselled front is the back view of the College from West North Sreet, where a dip in the ground displays to advantage the great mass of building, the Mitchell Hall with its great Gothic window, its angle-turrets and lofty buttresses.

The Northern Assurance Office stands at the angle between Union Street and Union Terrace. The clean surface and clear-cut lines of the granite masonry are very pleasant to the eye. Union Terrace contains some of the best modern buildings in the city—the Grand Hotel, the Aberdeen Savings Bank, which though very simple is an admirable specimen of a front specially designed for granite; the Offices of the Parish Council and the School Board, original and striking, the Public Library, the United Free South Church with its graceful dome, and His Majesty's Theatre—all serve to illustrate the changes that are being rung on granite fronts in recent years.

The contrast between these more ornate buildings and the severely classic simplicity of the Music Hall, a square block with a portico of Ionic pillars, belonging to the early nineteenth century, shows what a change in sentiment has taken place. The feature of all the Aberdeen architecture is the careful, conscientious workmanship, which always gives the impression of lasting solidity. The material is so irresponsive that without hard labour, no effect is produced.

Union Terrace and Gardens, before widening of Bridge

We can do no more than mention some of the other notable edifices in the city. The Grammar School, erected in 1863, is a successful application of castellated Gothic to a modern building—all the more effective that it

Grammar School, Aberdeen

is well set back from the street. The contiguous Art School and Art Gallery are modern buildings, each with an order of columns and a pediment which break the long low line of the facade. The elliptical arch that unites them gives access to Gordon's College, the centre portion of which is a piece of sober eighteenth century work. The wings and colonnades were added subsequently. The Head Office of the North of Scotland and Town and County Bank at the top of King Street has its entrance porch at the angle with a colonnade of pillars. Near it is the

Gordon's College, Aberdeen

Town Cross, a hexagonal erection with Ionic columns and a tapering shaft rising from the centre of the roof, with a heraldic unicorn as terminal. It dates from the end of the seventeenth century. In the panels of the balustrade are half-length portraits of Scottish and British Kings (including the seven Jameses). It is a fine example

Bridge of Don, from Balgownie

of its class and was the work of a local mason.   The
royal portraits are real and authentic.   The Ionic screen
or facade between Union Street and the city churches
gives some idea of the severely classic architecture that
was the vogue in Aberdeen nearly a century ago.

A word must be said about the chief bridges.   Union
Bridge has a span of 130 feet, and was built in 1802 to

Old Bridge of Dee, Aberdeen

facilitate the making of Union Street.   It was originally
narrower than the street and has recently been widened
to meet the requirements of increased traffic.   The
Bridge of Don (Balgownie), probably built early in the
fourteenth century if not earlier, throws its one Gothic arch
over the deep contracted stream of the river.   A small

bequest in the seventeenth century for its maintenance has been so well husbanded that out of its accumulations the cost of the new Bridge (£17,000), and other buildings has been defrayed, and the capital value of the fund— called the Bridge of Don fund—is to-day £26,500. The new bridge, much nearer the sea and with five arches, was designed by Telford and completed in 1830. The Old Bridge of Dee (with seven arches) was founded by Bishop Elphinstone and completed in 1527 by Bishop Gavin Dunbar. In 1842 it was widened 11½ feet. The New (Victoria) Bridge, a continuation of Market Street, was opened in 1882, since when quite a new and populous city has sprung up on the south side of the river, entirely eclipsing the old fishing village of Torry which formerly monopolised this side of the water.

21. Architecture—(d) Domestic.

The mansion-houses of the county, whether they are ancient fortalices modernised by later additions or entirely modern buildings erected within a century of the present time, deserve more space than can be allotted to them here. They are of all types of architecture, classical, renaissance, and composite, but there is no doubt that the castellated, Scotch baronial, the traditional type so common in the seventeenth century, still predominates.

Foremost among them must be mentioned Balmoral Castle far up the valley of the Dee. Built in 1853 of a light grey granite found in the neighbourhood, it is composed of two semi-detached squares with connecting wings, and displays the usual castellated towers, high-pitched gables and conical roofed turrets. The massive clock-tower rising to a height of 100 feet from amongst

Balmoral Castle

the surrounding leatage and gleaming white in summer sunshine forms a pleasing picture. The late Queen Victoria purchased the estate in 1848, and the Prince Consort took a great personal interest in the design the details of which are said to be modelled on a close study of Castle Fraser, already referred to. For more than half a century it has been a royal residence and though many additions and alterations have been made in that time, the general picture of the edifice remains the same to the traveller on the Deeside road. Two miles below is Abergeldie Castle, which has been leased by the Royal Family for many years. Its turreted square tower, old and plain and somewhat cramped in space, serves as a contrast to the more spacious modern mansion.

This region of the Dee has many mansions. Invercauld House, reconstructed in 1875, is in the same manner, its chief feature being a battlemented tower seventy feet high. The situation of Invercauld at the foot of a high hill and backed by plantations of pine and with a beautiful green terrace stretching to the river Dee is probably unsurpassed in the district. As seen from the Lion's Face Rock, a perpendicular cliff on the south side of the river, this house of the Farquharsons makes a striking picture not likely ever to be forgotten. Farther up is Mar Lodge, the residence of the Duke of Fife, in the horizontal and English domestic style. It was built so recently as 1898, and replaced a somewhat similar building destroyed by fire. Glenmuick House, built in 1873, is in the Tudor style, strongly treated and modified to harmonise with the rugged surroundings. The only other Deeside mansion we can refer to is Kincardine Lodge, recently built, a very fine building, based to a large extent on the plan of Fyvie Castle, which we have already referred to as the grandest castellated mansion-house in the north.

Donside is not so well furnished with stately and

luxurious manor-houses, but it has Castle Newe and Cluny
Castle, the antique-modern Place of Tilliefoure, Fintray
House in the Tudor style, Pitmathen in French Renais-

Cluny Castle

sance, each in its own way a work of art.   Midway
between the two valleys is Dunecht House, which was
built for Lord Lindsay, a great authority on Christian

art, and of which the most striking feature is the great campanile in the Italian manner.

In the Ythan valley, Haddo House, the residence of the Earl of Aberdeen, Lord Lieutenant of the County, belongs to the period of the late English Renaissance,

Haddo House

but additions have been made from time to time. Crimonmogate, Strichen, and Philorth are classic.

It is a curious fact, worthy of mention, that the local masons have almost developed a school of craftsmanship, by the thorough conscientiousness and downright honesty of their work. We have already remarked that Kintore and Inverurie seemed to be the centre from which the sculptured stones radiated. In the same region are the

Midmar Castle

group of castles, Castle Fraser, Craigievar, Midmar and Cluny (now destroyed), all within an easy radius of the centre. Castle Fraser and Midmar were built by a mason called John Bell, whose work was characterised by sterling qualities. The art would almost seem to have been handed down through several generations of craftsmen, for the modern Cluny Castle and Dunecht House, as well as their chapels, besides other palatial and extensive fabrics, were built entirely by local masons, without any extraneous help. It seems as if the building art were indigenous to this particular locality.

22. Communications—Roads, Railways.

In ancient times the chief means of communication between Aberdeenshire and the south was the old South and North Drove Road, which crosses the Cairn-o-Mounth from Fettercairn in Kincardine, and, passing the Dye and Whitestones on the Feugh, reaches the Dee at Potarch. It then ran along the hill to Lumphanan and on through Leochel to the Bridge of Alford, thence to Clatt and Kennethmont and along the valley of the Bogie to Huntly.

There was another—a supposed Roman road—which, coming up from the direction of Stonehaven, crossed the Dee at Peterculter, and, proceeding northward through Skene, Kinnellar, Kintore and Inverurie, went on to Pitcaple. Thence it passed through Rayne and across the east shoulder of Tillymorgan to what has been regarded as a Roman camp at Glenmailen, and by the Corse of Monellie, Lessendrum and Cobairdy, to the fords of the Deveron below Avochie.

Another ancient road crossed the mountains from

Blairgowrie by the Spittal of Glenshee, over the Cairnwell, Castleton of Braemar, and the upper waters of
the Gairn to the valley of the Avon at Inchrory and
thence by Tomintoul to Speyside.

After the '45 General Wade adopted the southern
part of this road as the line of his great military route

Spittal of Glenshee

from Blairgowrie to Fort George, but from Castleton
he turned to the east, went down the Dee valley to
Crathie, and thence across the hills to Corgarff in Upper
Strathdon from which he reached Tomintoul by the
"Lecht." This route he completed in 1750.

These roads had naturally to lead to fords in the rivers, and, when bridges came to be built, it was just as natural that they should be placed in the line of established routes. When the Bridge of Alford was built over the Don in 1810–11 and the Bridge of Potarch over the Dee in 1812–13, a new line of road was made across country to connect them. It went from Dess through Lumphanan and Leochel to the Don valley.

The first turnpike made in Aberdeenshire was the road from the Bridge of Dee to the city of Aberdeen *via* Holborn Street, which completed the northern section of the great post-road between Edinburgh and Aberdeen. This was in 1796.

About the same time was made the North Deeside Road reaching from Aberdeen to Aboyne and thence to Ballater, Crathie, and Braemar, where it met the Cairnwell Road. Another was the Aberdeen and Tarland route, which went by Skene and Echt with branches joining on to those already in existence. One of these struck off at Skene, and, crossing the hill of Tilliefourie, proceeded to Alford. It was afterwards extended up the Strath by Mossat, and Glenkindie to Corgarff, where it met General Wade's road.

The great post-road from Aberdeen to Inverness went by Woodside, Bucksburn, Kintore, Inverurie, the Glens of Foudland to Huntly and Cairnie on the boundary of Banffshire. It had branches from Huntly to Portsoy through Rothiemay and to Banff through Forgue by the Bridge of Marnoch.

The Strathbogie Road from Huntly to Donside by

way of Gartly, Rhynie, and Lumsden joined the Strathdon Road at Mossat. Though by no means the most
convenient, it is still used as the route along which the
mails are conveyed to Strathdon.

The Aberdeen and Banff Road left the post-road at
Bucksburn and passing through Dyce, New Machar, Old
Meldrum, Fyvie, Turriff, and King Edward made for the
Bridge of Banff.

In the eastern district the most important route was
that to Peterhead. It crossed the new Bridge of Don,
and, passing through Belhelvie, Ellon, and Cruden, came
to Peterhead by the coast. From there it went straight
across country to Banff by Longside, Mintlaw, New
Pitsligo, and Byth, thence over the Longmanhill to
Macduff. Later a coast route was made connecting
Peterhead and Fraserburgh, by way of St Fergus, Crimond and Lonmay. Another continuation of it was along
the coast past Rosehearty, Pennan, Gardenstown and Troup
Head into Banffshire.

It was only during the nineteenth century that proper
and serviceable highways were constructed. Prior to that
time a few main roads had been made but side connections
were few and badly kept, so that wheeled vehicles, if they
had existed, would have been a useless luxury. Early in
the eighteenth century wheeled vehicles were absolutely
unknown. In 1765 the judges of the Circuit Court of
Justiciary first travelled to Aberdeen in chaises instead
of on horseback. The first mail coach did not arrive
till 1798. It took 21 hours between Edinburgh and
Aberdeen. Not till 1811 did passenger coaches begin

to ply between Aberdeen and Huntly. Then only was it possible for the farmer to convey his products by cart, which superseded the pack-horse as a means of transport.

The upkeep of the roads was secured by a system of tolls. Traces of the system still survive in the renovated toll-bar houses, which in some cases retain a window facing right and a window facing left to mark the approach of vehicles from either side. Aberdeenshire abolished tolls in 1865.

The Railway system reached Aberdeen in 1848. Prior to that time for fifty years the stage coach plying between Edinburgh and Aberdeen had been, apart from the sea-routes, the only bond between this part of the country and the south. A few years later, in 1854, what is now the Great North of Scotland Railway was opened from Aberdeen to Huntly, and two years thereafter was extended as far as Keith. This is still the main line of railway in the county. It touches in its course Dyce, Kintore, Inverurie, Insch, and Huntly. By and by branch lines were constructed forking off from it at various points; *first* from Inveramsay, through Wartle, Fyvie, to Turriff and ultimately to Macduff; *second* from Inverurie across country to Old Meldrum; *third* from Kintore up Donside by Kenmay and Monymusk to Alford; and lastly from Dyce through New Machar, Udny, Ellon to New Maud, where it bifurcates, one fork going on to Peterhead the other to Fraserburgh. This is the Buchan and Formartin branch. Recently a sub-branch was made from Ellon running to the coast and touching Cruden

Bay, its terminus being at Boddam within half an hour's distance of Peterhead. From Fraserburgh, a light railway runs to Cairnbulg, Inverallochy and St Combs. The only other line of railway in the county is the Deeside line, which runs up the Dee valley as far as Ballater. It was begun in 1853, and Banchory was the terminus till 1859, when an extension was made to Aboyne; then in 1866 it was extended to Ballater.

The lack of population and the paucity of goods apart from agricultural products have handicapped the local railways, which are far from prosperous. The chances of extension in other directions are very remote. Meantime outlying districts, such as Strathdon and Braemar, are served by motors. The holiday and tourist traffic during the summer months and the influx of sportsmen at the shooting season are contributory sources of revenue, but even these show no tendency to grow—a state of affairs due to the prevalent use of private motor-cars.

Aberdeenshire has no canals and is never likely to have. Prior to the advent of railways a canal, designed by Telford, the great engineer, was constructed between Aberdeen and Port Elphinstone on the south side of Inverurie. It was opened for passenger and goods traffic in 1806, and continued to serve the district until the steam-engine sounded its knell. For nearly half a century it was a bond between the chief city and the centre of the county and, although it never was remunerative to the promoters, and provided a very slow mode of conveyance, it was of great public service. The railway line to the north runs parallel at certain places to the track of this canal, whose superannuated embankments may still be recognised, after half a century, at various points between Aberdeen and Inverurie.

23. Administration and Divisions.

In the twelfth century Scotland was divided into Sheriffdoms, where the Sheriff was the minister of the Crown for trying civil and criminal cases. The office was hereditary until the rebellion of '45, when its hereditary character was abolished. Aberdeenshire has a non-resident Sheriff-Principal (who is also Sheriff of Banff and Kincardine) besides two resident Sheriff-substitutes. These deal with ordinary civil cases such as debts, as well as with criminal cases involving fine or imprisonment, but not as a rule involving penal servitude, except forgery, robbery and fire-raising.

The head of the county is the Lord-Lieutenant. Next to him is the Vice-Lieutenant and a large number of Deputy-Lieutenants and Justices of the Peace, but the chief administrative body is the County Council, which consists of 65 members. These elect the chairman and vice-chairman, who are designated respectively convener and vice-convener. County Councils were first established in 1889. The county is divided into districts, and each district has so many divisions, or parishes, which elect one councillor. Aberdeenshire has 85 parishes, which are grouped in eight districts : (1) Deer with fifteen electoral divisions, (2) Ellon with seven, (3) Garioch with six, (4)

Deeside with six, (5) Turriff with seven, (6) Aberdeen with nine, (7) Alford with four, and (8) Huntly with four, making fifty-eight electoral divisions in all. The powers of the Council are to maintain roads and bridges, to administer the Contagious Diseases (Animals) Acts, to appoint a medical officer of health and a sanitary inspector, to deal with the pollution of rivers and to see to the protection of wild birds. Previous to the passing of the Act of 1889 the Commissioners of Supply were the chief governing body. They are still retained but have no jurisdiction, except in so far as they elect members to the Standing Joint Committee. This committee includes representatives from the County Council appointed annually and from the Commissioners of Supply, together with the Sheriff *ex officio*. The Standing Joint Committee has charge of the Police and controls all the capital expenditure in the county.

Each district has a district committee consisting of the county councillors for the divisions of the district and of parish councillors selected by each parish council of the district. Each parish has in this way two representatives on the district committee, one elected by the electors and the other appointed by the parish council. This district committee is the local authority for administering the Public Health Acts, but has no power to raise money—that being the function of the County Council as a whole.

By a later Act of 1894, a parish council was established in every parish. The number of councillors in landward parishes is fixed by the County Council and

in burghal parishes by the Town Council. The parish
council looks after the Poor Law and must provide for
pauper lunatics, sees to the levying of the school rate,
to the administration of the Vaccination Acts, and to the
appointment of Registrars.

The affairs of the county are therefore divided amongst
three bodies, the County Council, the District Committees
and the Parish Councils. Prior to 1890 the powers of
local administration lay with the Commissioners of Supply,
the Road Trustees and the Parochial Boards, whose func-
tions are now vested in these other bodies.

Each parish, besides having a Parish Council, has a
School Board, which, since 1872, has administered the
education of the parish. Education is free and compul-
sory for all children between the ages of 5 and 14.
The schools are of three types—primary, intermediate,
and secondary. The intermediate schools provide a
three years' course beyond the elementary stage, and the
secondary schools a further course lasting for two years.

The County Council now takes a certain share in
educational administration, having powers to allocate
grants of money to schools and bursaries to pupils. The
training of teachers, which until recently was in the hands
of the Churches (Established and United Free), has now
passed to a Provincial Committee elected by various representative bodies.

Every burgh has a Town Council consisting of Provost, Magistrates and Councillors, who hold their seats for
three years. The number of councillors varies with the
size of the town. In Aberdeen, the councillors are elected

by wards, of which there are eleven, each ward electing three representatives, one of whom retires annually. The Town Council of Aberdeen consists of 34 members, the Dean of Guild being an *ex officio* member. The Town Council is the local authority for Public Health, and looks after the streets, buildings and sewers. It owns the gas works, water works, tramways, electric power station, and public parks. It regulates the lighting, cleansing, and sanitation. The Magistrates, who are elected annually by the Council, are the licensing authority, and form the police court for the trial of minor offences.

The city of Aberdeen is not like Peterhead and Fraserburgh included in the administration of the county, being itself constituted the county of a city, with a Lord-Lieutenant of its own, who is the Lord-Provost *ex officio*. It has its own Parish Council as well as its own School Board.

Aberdeenshire is represented in Parliament by four members—two for the county, east and west, and two for the city, north and south. Some of the smaller burghs, Kintore, Inverurie and Peterhead, are grouped with similar burghs in Banff and Moray (Banff, Elgin, Cullen) to form a constituency called the Elgin Burghs, which returns one member. In addition, the University of Aberdeen shares a member with the University of Glasgow.

There is still a certain amount of overlapping and confusion in the administrative divisions. For example, Torry, which is on the Kincardineshire side of the Dee, is really a suburb of Aberdeen, and as such elects members to the Town Council, the Parish Council, and the School Board, but it has no share in electing a member of Parliament for Aberdeen, being in that regard part of Kincardineshire, and voting for a representative of that county. There are other similar anomalies.

24. The Roll of Honour.

It is an accepted fact that Aberdonians have intellectual characteristics somewhat different from those of their fellow-countrymen, the result partly of race, partly and chiefly, we believe, of environment. We have already alluded to the amalgamation of nationalities that went to form the people of this north-eastern corner of the kingdom. Doubtless the Spartan upbringing that was the rule in the county served to develop sturdy character and good physique. The result is that the Aberdonian has distinguished himself in all parts of the Empire and even beyond it. Not that he has often risen to the front rank of greatness, but he is frequently found well forward among the best of the second-class.

Their own county presenting no tempting openings for ability, Aberdonians have migrated from the narrow home-sphere in great numbers and have made their mark as administrators, medical officers, and even as soldiers of fortune. In the seventeenth and eighteenth centuries the cadets of the great houses, exiled by the pressure of the times, joined the service of continental kings and rose to high rank in the armies of Sweden, France, and Russia.

Chief amongst these was James Keith, younger brother of the last Earl Marischal, and born at Inverugie Castle. After serving for nineteen years in Russia, he joined the service of Frederick the Great, under whom he attained to the highest military rank as Field-Marshal, contributing to victories gained during the Seven Years' War and conducting the retreat from Olmütz. At the battle of Hochkirchen, when charging the enemy, he fell mortally wounded in 1758. Peterhead keeps his memory green by a statue presented to it by the Emperor William I. It is a replica in bronze of a similar effigy in Berlin. Field-Marshal Keith is probably the native of Aberdeenshire who has figured most largely in history. He was Frederick's right hand, and his military genius has been fittingly acknowledged by Carlyle in his great work.

Another of the same type, though less eminent, was Patrick Gordon of Auchleuchries, who fought both on the Swedish and on the Polish side, but ultimately transferred his sword to Russia, where he rose to the highest rank, and on his death-bed was watched over and wept over by Peter the Great. He was born in 1635 at Auchleuchries near Ellon and died in 1699. He was a perfect example of the successful military adventurer, one of the type so skilfully depicted by Walter Scott in Dugald Dalgetty.

The county has been a prolific recruiting ground for the Army. After the '45 Chatham's device for breaking down the clan system and diverting the energies of the Highlanders into healthier channels by enlisting them in British regiments was an inspiration of genius. In 1794 the Duke of Gordon raised during a few weeks a regiment of Gordon Highlanders, which first distinguished itself with Sir Ralph Abercromby in Egypt, and did noble service also in the Peninsula and at Waterloo.

In the work of empire-making in India and elsewhere, the Aberdonian has borne a notable part. He has shown ability to exercise a singular mastery over inferior races. Conspicuous in this respect was Sir Harry B. Lumsden, who formed the Corps of Guides out of the most daring free-booters of the North-West frontier of India.

In statesmanship the county has been surpassed by other districts, and yet it has the distinction of having produced one Prime Minister—the fourth Earl of Aberbeen (1784–1860), who was responsible for the Crimean War, and whom Byron styled "the travelled Thane, Athenian Aberdeen."

The ecclesiasts of distinction are too numerous to mention. Foremost amongst them was Bishop Elphinstone, who, though not a native of the county, identified himself with its interests when he became Bishop (1483), founded the University, King's College, the light of the North (1494), and the church of St Machar (the Cathedral in Old Aberdeen) and was a pioneer in all that makes for educational enlightenment. He was instrumental in introducing the art of printing into Scotland. His tomb is very appropriately in King's College, the centre from which radiated the beneficent influence of his life. Henry Scougal (1650–1678), scholar and saint, son of Bishop Scougal and the inspirer of John Wesley, was a student of King's College. He had not been long ordained in his charge at Auchterless before he was appointed to the Chair of Divinity in King's College. He died at 28; but his *Life of God in the Soul of Man* is still greatly prized by lovers of devotional literature. Dean Ramsay, whose *Reminiscences of Scottish Life and Character* (1858) is a classic in humorous literature and one not likely soon to be forgotten, was born in Aberdeen.

In medical science the roll of eminent names is long and impressive, from Bannerman, who was physician to David II, down to Arthur Johnston, who after an academic career abroad, cultivated the muses at Aberdeen, gaining fame as a writer of Latin verse. He was for some time physician to Charles I. Born at Caskieben in 1587, he was rector of King's College in 1637, and died in 1641. Dr John Arbuthnot, though a native of Kincardineshire, was a student at Marischal College; as the friend of Pope and Swift, and the wit and physician at the Court of Queen Anne, he is likely to be remembered. Another celebrated physician was Dr John Abercrombie, who, born in Aberdeen, went to Edinburgh, and became head of the profession and first physician to the king in 1824. Others no less noted were Sir James Clark; Sir Andrew Clark; Neil Arnott, a contemporary with Byron at the Grammar School, and more famous as natural philosopher than as physician, devising skilful inventions in healing and ventilation; Sir James Macgrigor, to whose memory a lofty obelisk in polished red granite was erected in Marischal College quadrangle. After standing there for years it was recently removed to the Duthie Park. Macgrigor was a pioneer in the humanitarian treatment of the sick and wounded in war, and was chief of the Medical Staff in the Peninsular campaigns.

In natural science William Macgillivray is known by his careful and authoritative work on the *History of British Birds*. James Clerk Maxwell, who did so much for the advancement of modern Physics, was for a few years professor of Natural Philosophy in Marischal College. Dr Alexander Forsyth, minister of Belhelvie, invented the percussion lock, and Patrick Ferguson, a native of Pitfour, invented the breech-loading rifle.

The county is remarkable for families with pronounced hereditary intellectual gifts. The most noted case is that of the Gregories, who sprang from John Gregory, minister of Drumoak. It has produced fourteen Professors in British universities, skilled in Mathematics, Astronomy, Chemistry and Medicine. One of them was the inventor of the reflecting telescope. The Reids, the Fordyces, the Johnstons are other cases less remarkable, but still exceptional.

Philosophy is a sphere in which the Aberdonian has left his mark. The greatest local name in this regard is that of Thomas Reid, who created the Scottish school in opposition to David Hume, and whose *Inquiry into the Human Mind on the principles of Common Sense* was written while he was a Professor at King's College. Born at Strachan on the south side of the Dee, he was for a time parish minister of New Machar. Later he migrated in 1763 to Glasgow, as successor to Adam Smith. His *Intellectual and Active Powers* was written after his retirement in 1780. Other philosophical writers worthy of mention are, Dr George Campbell, who, besides his dissertation on *Miracles*, wrote a *Philosophy of*

Rhetoric ; Dr James Beattie, whose *Minstrel* is still read and whose *Essay on Truth* had a great contemporary reputation ; Dr Alexander Bain, an analytical psychologist, whose books *The Senses and the Intellect* and *The Emotions and the Will* contain the most complete ana-

Professor Thomas Reid, D.D.

lytical exposition of the mind.   Bain was the first Professor of Logic at Aberdeen, and in conjunction with his pupil Croom Robertson started the philosophical Review called *Mind*.

The sphere of imaginative literature is not the Aberdonian's sphere. Criticism, Philosophy, History, Science are more in his way, and yet a few names can be given as of some note in pure literature. Foremost in time and unrivalled in his own department is Barbour, Archdeacon of Aberdeen (1357). He studied at Oxford, and was contemporary with Wycliffe and Chaucer. His great work is *The Brus*, the most national of all Scottish poems. It is instinct with the spirit of freedom, of chivalry and romance, and details the struggles, the perils, and the marvellous escapes of his hero Robert the Bruce, with great simplicity, vividness, and directness. Alongside of him we may place John Skinner, author of *Tullochgorum*, *The Ewie wi' the Crookit Horn*, and other well-known songs. A native of Birse and for long episcopal minister at Longside, he was the father of Bishop Skinner. His fame rests on *Tullochgorum*, which Burns pronounced to be the best of Scotch songs. Dr W. C. Smith, the author of *Olrig Grange* and *Borland Hall*, was born and educated in Aberdeen. Dr George Macdonald, poet, novelist, and critic, author of *Alec Forbes* and other novels embodying local colour and illustrating Aberdeenshire life and dialect in the early part of last century, was a native of Huntly. The best known poet connected with Aberdeen is Byron, who spent some years of his boyhood in the city and short periods of the summer on Deeside. These visits to Ballater are reflected in his poem on Loch-na-gar and elsewhere in his work. He left Aberdeen at the age of 10 in 1798 and never saw it again.

History is a subject that has appealed to Aberdonians. Dr David Masson's monumental work on Milton must be mentioned. Other historians are Joseph Robertson,

John Stuart, John Hill Burton, Bishop Burnet, who wrote the *History of his Own Time*, Sir John Skene, and Robert Gordon of Straloch, antiquarian and map-maker, as well as his son James, minister of Rothiemay and historian of the early years of the Troubles. The first Principal of the University, Hector Boece, wrote histories somewhat credulous and imaginative but quite authoritative where his own times are concerned.

Of painters connected with the district may be mentioned Jamesone, Dyce, and Phillip called "of Spain" from his success with Spanish subjects. Architecture claims Gibbs, whose Radcliffe Library at Oxford, and London churches such as St Martin's-in-the-Fields, still stand a testimony to his art; the Smiths and Archibald Simpson have already been mentioned. Sculpture owns the two Brodies and Sir John Steell.

Scholars like Wedderburn and Ruddiman, Cruden of *Concordance* reputation, Dr James Legge, the Orientalist, and Professor Robertson Smith, born in the Donside parish of Keig, editor of the *Encyclopaedia Britannica*, Professor of Arabic at Cambridge, and one of the most learned pundits of his time—are but a few representatives of a long list.

The thirst for education and the well-taught parish schools of the county contributed to bring about such results. The doors of the University have for centuries been opened by bursaries to the poorest boys, and in this way many who were endowed with capacity above ordinary entered the learned professions and rose to eminence.

25. THE CHIEF TOWNS AND VILLAGES OF ABERDEENSHIRE.

(The figures in brackets after each name give the population in
1911, and those at the end of each section are references
to pages in the text.)

Aberdeen (161,952).   From being entirely built of granite,
Aberdeen is best known as "The Granite City." The light
grey stone gives the town a clean look which strikes visitors
from cities built of brick or of sandstone. Its many handsome
public buildings, banks, offices, churches and schools, all solid
and substantial, and of great architectural interest, are undoubtedly finer than those of any other town of the same size
in the kingdom.

The first historical reference to it is in the twelfth century;
later a charter was obtained from King William the Lion,
granting the city certain trading privileges. Long before Edinburgh and Glasgow had begun to show signs of rising to greatness,
Aberdeen was a port of extensive trade, but its growth was slow
until the dawn of the nineteenth century. In 1801, its population
was only 27,608; in 1831 this figure had doubled, and in recent
years, owing chiefly to the phenomenal growth of the fishing
industry, its progress has been rapid.

Aberdeen has long been a great educational centre. Its
Grammar School claims to have existed in the thirteenth century.

Its first University, King's College in Old Aberdeen, was founded in 1494 by Bishop Elphinstone, and its second, Marischal College in New Aberdeen, by Earl Marischal in 1593. These were united in 1860 as the University of Aberdeen. Since that time the buildings of both Colleges have been largely added to, and the number of professorships greatly increased. Its students in the different faculties, Arts, Medicine, Science, Law and Divinity are little short of 1000.

Being the only really large town in the county, and for that matter in the whole north of Scotland, it tends to grow in

The Old Grammar School, Schoolhill

importance, and its business connections are ever extending. It is the focus of the trawling industry, and of the granite trade; while the agricultural interests of the county look to Aberdeen as their chief mart and distributing medium. Its secondary schools, its technical college, its agricultural college, its University, all help to swell its population by bringing strangers to reside within its boundaries. In itself it is clean, healthy and attractively built, while its fairly equable climate, its relatively low rain-fall (29 inches) and its equally low death-rate (14·2 per 1000) conduce to its popularity as a residential town. Being the northern terminus of the Caledonian Railway, and having excellent service to London by the West Coast, the Midland, and the East Coast routes, it obtains a large share of the tourist traffic; and the sportsmen who fish in the Aberdeenshire rivers or shoot grouse in the Aberdeenshire moors must all do more or less homage to the county town.

The chief street of the city is Union Street created a century ago at a cost which was considered reckless at the time but which has been more than justified by the results. This first improvement scheme, which has been followed up by others in recent times, was the work of men with a wide outlook. Prominent among the Provosts of enlightenment was Sir Alexander Anderson, whose name is now at the eleventh hour stamped in memory by the Anderson Drive—a fashionable west-end thoroughfare. Union Street is the backbone from which all the other thoroughfares radiate. It is broad and handsome and the buildings that face each other across it are as a rule worthy of the street. Union Bridge, one of Fletcher's graceful structures, with a span of 130 feet, makes a pleasing break in the line of buildings and permits a view north and south along the Denburn valley. The northern view, which shows Union Terrace and Union Terrace Gardens with handsome public buildings, both in the foreground and in the background, is undoubtedly one of the finest in the city. The Duthie Park on the north bank of the Dee, the links that fringe the northern coast, the picturesquely wooded amenities of Donside, above and below Balgownie Bridge, the quaint other world air of Old Aberdeen with its lofty trees, its grand cathedral and the ancient crown of King's College, these are all elevating and meliorating influences that help to keep in check the commercial spirit that rules about the harbour-quays and the fish-market.

Aberdeen can boast of four daily newspapers besides several weeklies. It claims the honour of having the oldest newspaper in

Scotland—*The Aberdeen Journal*—established in 1748. (pp. 3, 8, 11, 13, 20, 24, 37, 38, 39, 66, 68, 75, 80, 83, 85, 89, 91, 92, 94, 95, 101, 102, 107, 108, 109, 111, 126, 145, 162, 164, 165, 169, 172, 173.)

Aberdour (549) is a small village on the coast half-way between Troup Head and Rosehearty. Sometimes called New Aberdour to distinguish it from the parish, the village came into existence in 1796. The parish is very ancient. Its church, now in ruins, was dedicated to St Drostan, the disciple and companion of St Columba. Aberdour is the birth-place of Dr Andrew Findlater, once Head-master of Gordon's Hospital (now College), Aberdeen, and first editor of Chambers's *Encyclopaedia*. (pp. 39, 62, 105, 119, 164.)

Aboyne (1525), properly called Charlestown of Aboyne in compliment to the first Earl of Aboyne, is a picturesquely situated village on Deeside with a high reputation for its bracing climate. Near it is Aboyne Castle—for centuries the family seat of the Marquis of Huntly. In the vicinity are Lochs Kinnord and Davan. At Dinnet are beds of kieselguhr. (pp. 2, 8, 24, 31, 88, 117, 119, 162.)

Alford (pa. 1464), on Donside, is the terminus of the branch railway from Kintore and the centre of a rich agricultural district called the Vale of Alford. In the neighbourhood are several interesting castles—Terpersie, Kildrummy and Craigievar. From Alford the main Donside road leads up the valley to Strathdon and Corgarff, from which there are passes both to Deeside and to Speyside. (pp. 27, 71, 113, 115, 134, 160, 162.)

Ballater (1240), a small town beautifully situated on the north side of the river Dee, in a level space enclosed by high mountains, is 660 feet above sea-level. From Ballater coaches drive daily to Braemar, passing Balmoral Castle half-way. (pp. 2, 8, 18, 24, 27, 33, 154, 162, 164, 176.)

Birsemore Loch and Craigendinnie, Aboyne

Braemar (502), properly Castleton of Braemar, is the highest village in the county, being 1100 feet above sea-level. It stands at the junction of the Clunie and the Dee, and is finely

Mar Castle

sheltered in a hollow amongst the surrounding mountains. Braemar is a fashionable health resort. Some 10,000 strangers visit it annually. At the beginning of the nineteenth century it was not much more than a Highland clachan. Now it has

Ballater, view from Pannanich

spacious hotels with electric light and all modern conveniences on a luxurious scale. Six miles distant is the famous Linn of Dee. The Duke of Fife's Highland residence, Mar Lodge, as well as Mar Castle and Invercauld House, the home of the Farquharsons, are all in the vicinity. From Braemar the ascent of Ben-Macdhui is usually made, and sometimes also Loch-na-gar. A road leads from Braemar up the valley of the Clunie and over the Cairnwell to Blairgowrie. (pp. 8, 21, 22, 33, 66, 68, 75, 112, 161, 162.)

Byth (360), usually New Byth, is a village three miles from Cuminestown, and founded in 1764. It is a bare and treeless district. Near it are the hills of Fishrie with a large number of crofts given off by the Earl of Fife in 1830 to poor people evicted from other estates at a time when the fashion began of amalgamating small holdings in larger farms. (pp. 91, 124, 163.)

Collieston is a fishing village circling round a romantic bay near the parish church of Slains. Here in 1588 one of the ships of the Spanish Armada (*Santa Catherina*) was wrecked. The fishermen still call the creek St Catherine's Dub. Several small cannon have been recovered from the pool. Eighty years ago, Collieston enjoyed a certain notoriety for smuggling, and the graveyard of Slains close by contains evidence of the deeds of violence that the contraband trade brought about. (pp. 38, 48, 54.)

Culter, eight miles west of Aberdeen, celebrated for its paper-mills, which date back to 1750. This paper-mill, the first of its kind in the north, manufactured superfine paper and in particular the bank-notes of the Aberdeen Bank. (p. 89.)

Cuminestown (466), a village on the north side of the Waggle Hill in Monquhitter, was established by Joseph Cumine of Auchry in 1763. Joseph Cumine was a pioneer in agricultural improvement. He planted trees and started the manufacture of linen. About a mile distant is the smaller village of Garmond. The villages were once much more populous

Braemar from Craig Coynach

in the days when the spinning of flax and the knitting of stockings were rural industries.   (p. 40.)

Ellon (1307), a thriving town on the Ythan, is the junction for the Cruden and Boddam Railway.   It has a shoe factory and large auction marts for the sale of cattle.   The Episcopal Church —St Mary's on the Rock—was designed by George Edmund Street and is a handsome building in Early English style.   A prominent divine in the pre-Disruption controversies, Dr James Robertson, was parish minister of Ellon from 1832 to 1843.   Later he became a professor in Edinburgh University.   Ellon is a place of great antiquity.   It was the seat of jurisdiction of the Earldom of Buchan, and there the earls held their Head Court. (pp. 29, 41, 163, 164, 171.)

Fraserburgh (10,570) is the third largest town in the county. It is a busy, thriving place, being the great centre of the herring fishing industry in Scotland.   It was founded by Sir Alexander Fraser, one of the Frasers of Philorth (now represented by Lord Saltoun).   The Frasers are said to have come into England with the Normans.   A royal charter was granted in 1546 erecting "Faithlie" as it was then called into a free burgh of barony with all the privileges.   Sir Alexander Fraser was a great favourite with James VI and was knighted at the baptism of Prince Henry, 1594: he was a man of enterprise; he built the town and the harbour and erected public buildings.   He received from King James the privilege of founding a University in Fraserburgh, and a building was set apart for this institution.   Not only so but a Principal was appointed in 1600.   The College may have been active for a few years, but very little is known of its history. During the plague which raged for two years at Aberdeen, the students of King's College went for safety to Fraserburgh in 1647 and, it is supposed, occupied the old College buildings.   A street in the town is still called "College Bounds."   (pp. 8, 38, 59, 61, 76, 94, 95, 101, 102, 163, 164, 169.)

The Doorway, Huntly Castle

Huntly (4229), the largest inland town of the county, is situated at the confluence of the Deveron and the Bogie. It is the centre of an extensive agricultural district—Strathbogie—and has woollen and other manufactures. In the vicinity are the ruins of Huntly Castle, the property of the Duke of Richmond

The Bass, Inverurie

and Gordon. The first Lords of Strathbogie, being opposed to Bruce's claims of kingship, were disinherited and their lands bestowed on Sir Adam Gordon, whose descendants became Earls of Huntly, Marquises of Huntly and Dukes of Gordon. The old castle of Strathbogie was destroyed after the battle of Glenlivat in 1594, but rebuilt as Huntly Castle in 1602. Huntly is the birth-place of Dr George Macdonald, poet and novelist. (pp. 29, 31, 91, 160, 162, 164, 176.)

Insch (616), a village on the Great North Railway, with Benachie on one side, and the Culsalmond and Foudland Hills on the other. The vitrified fort of Dunnideer is in the vicinity. (pp. 119, 164.)

Inverurie (4069), a royal burgh at the confluence of the Ury and the Don. The workshops of the Great North of Scotland Railway were removed from Kittybrewster to Inverurie some years ago, thereby increasing the population of the burgh. It is one of the Elgin parliamentary burghs. The Bass of Inverurie is a conical mound, long considered artificial, but now ascertained to be a natural formation due to the action of the two rivers. Inverurie has paper manufactures. In the neighbourhood is Keith Hall, the seat of the Earl of Kintore. (pp. 27, 71, 80, 108, 109, 113, 132, 137, 160, 162, 164, 165, 169.)

Kemnay (948), about five miles up Donside from Kintore, is well known for its extensive granite quarries, which sent stones to build the Forth Bridge and the Thames Embankment. Near it is Castle Fraser, one of the finest inhabited castles of the county. Fetternear, once the county seat of the bishops of Aberdeen, is on the opposite side of the river. (p. 83.)

Kintore (818) is a royal burgh of great antiquity. A mile to the west are the ruins of Hallforest, destroyed in 1639. Kintore has, in its vicinity, several "Druidical" circles and sculptured stones. (pp. 27, 49, 108, 136, 160, 162, 164, 169.)

Longside (392) dates from 1801. A woollen factory brought for a time prosperity to the village, but this has been given up and the population dwindles. Rev. John Skinner, the author of *Tullochgorum*, was for over sixty years minister of the Episcopal Church at Linshart, close to the village of Longside. Here also was born Jamie Fleeman, "the laird of Udny's fool," a half-witted person whose blunt outspoken manner and shrewd remarks are still widely remembered. (pp. 163, 176.)

Maud is the point where the Buchan railway bifurcates for Peterhead and Fraserburgh. Maud is a centre for auction sales of cattle. (p. 164.)

Mintlaw (377) was founded about the same period as Long-side and the fortunes of both villages, which are three miles apart, have been similar. (p. 163.)

Newburgh (537), on the estuary of the Ythan, was at one time notorious like Collieston for smuggling. Ships of small burden still come up to its wharf at full tide and sometimes proceed as far as Waterton. The bed of the estuary of the Ythan is covered with mussels, much used in the past as bait by the local fishermen, as well as for export to other fishing stations. The revenue from this source has greatly fallen off in recent years—line-fishing having suffered from the rise of trawling. (p. 48.)

New Deer (675) is a village established about 1805. Brucklay Castle, the seat of the Dingwall-Fordyce family, recently converted into a mansion of the old Scottish castellated style, and surrounded with tasteful grounds, is now one of the most charming edifices in the district. A mile to the west is the ruined castle of Fedderat. (p. 138.)

New Pitsligo (pa. 2226) is a village in the neighbourhood of the sources of the Ugie, and extending for a mile in two parallel streets along the eastern slope of the hill of Turlundie. It stands 500 feet above sea-level. The village takes its name from Sir William Forbes of Pitsligo, who founded it in 1787. Here a linen trade was at one time carried on; this gave place to hand-loom weaving and ultimately to lace-making. The Episcopal Church was designed by G. Edmund Street, and it is said to be one of the best examples of his work in Scotland. The manufacture of moss-litter from the peat in the neighbourhood was recently started. (pp. 91, 163.)

Old Deer (179) is prettily situated on the South Ugie. The district has memories of St Columba and St Drostan. In the neighbourhood are the ruins of a Cistercian Abbey, and "Druidical" circles. (pp. 2, 105, 113, 115, 117, 123.)

Old Meldrum (1110) was erected by charter into a burgh of barony in 1672. It is well known for its turnip-seed. It used to employ many persons in handloom weaving and in the knitting of stockings. Both industries have fallen to decay, and the population tends to dwindle. There is a long-established distillery in the town. (pp. 108, 112, 163, 164.)

Peterhead (13,560), the most easterly town in Scotland, is built of red granite. A century ago it was a fashionable watering-place, and used to be a whaling station. Now its chief industry is the herring fishing. South of the town a harbour of refuge is being constructed by convict labour, from the convict prison close by. The harbour of refuge will cost, it is said, a million of money and its construction will occupy 25 to 30 years. A linen factory once existed here, as also a woollen factory, which exported cloth to the value of £12,000 a year. Both became extinct, but the woollen industry was revived and still prospers. Another prominent industry is granite polishing. At Inverugie Castle was born Field-Marshal James Keith, whose statue stands in front of the Town-House. The "Pretender" landed at Peterhead on Christmas Day, 1715. Peterhead was erected into a burgh of barony in 1593 by Earl Marischal, the founder of Marischal College, Aberdeen. It continued to be part of the Earl's estates till the rebellion of 1715, when the lands were confiscated. The Peterhead portion is now the property of the Merchant Maiden Hospital of Edinburgh. (pp. 8, 29, 38, 39, 41, 57, 66, 68, 76, 83, 94, 95, 101, 102, 111, 163, 164, 169, 171.)

Rosehearty (1308) is a misspelt Gaelic name of which *Ros*, a promontory, and *ard*, a height, are undoubted elements. The little town stands on the shore a mile north of Pitsligo[1]. There is a tradition that in the fourteenth century a party of Danes landed and took up residence here, instructing the inhabitants, who were mostly crofters, in the art of fishing. (pp. 62, 162.)

The White Horse on Mormond Hill

[1] Alexander Forbes, fourth and last Lord Pitsligo (1678—1762) was a warm supporter of the exiled Stuarts and took part in both rebellions. After Culloden, he remained in hiding, his chief place of concealment being a cave in the rocks west of Rosehearty.

Strichen (1094) was formerly called Mormond, from the hill at the base of which the village stands. This hill owing to the comparatively level character of the surrounding country is a conspicuous feature in the landscape for miles. On the south-western side, the figure of a horse is cut out in the turf, the space being filled up with white stones. This "White Horse" occupies half an acre of ground and is visible at a great distance. On the south side of the hill an antlered stag on a larger scale is figured in the same manner. This was done so late as 1870. (pp. 11, 16, 38, 91, 158.)

Torphins (455), a rising village on Deeside, much resorted to by Aberdonians in the summer months. (p. 106.)

Turriff (2346) is situated on a table-land on the north of the burn of Turriff near its junction with the Deveron. Turriff is midway between Aberdeen and Elgin; hence the couplet—

> Choose ye, choise ye, at the Cross o' Turra
> Either gang to Aberdeen or Elgin o' Moray.

Turriff is very ancient, being mentioned in the *Book of Deer*, under the name of Turbruad, as the seat of a Celtic monastery dedicated to St Congan, a follower of St Columba. The double belfry of the old church (date 1635) is really a piece of castellated architecture applied to an ecclesiastical edifice. The churchyard gateway is also Early Scottish Renaissance. (pp. 11, 31, 40, 110, 112, 119, 163, 164.)

Fig. 1. Area of Aberdeen-
shire compared with that
of Scotland

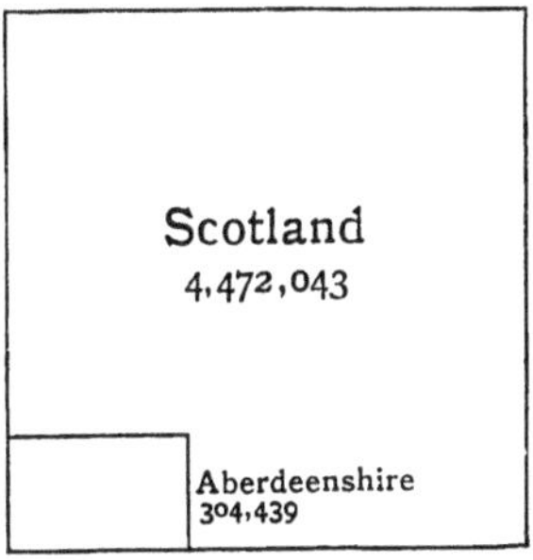

Fig. 2. Population of Aberdeenshire compared with
that of Scotland

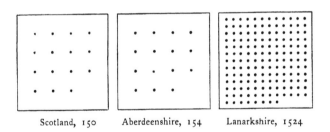

Scotland, 150   Aberdeenshire, 154   Lanarkshire, 1524

Fig. 3. Comparative density of Population to square mile

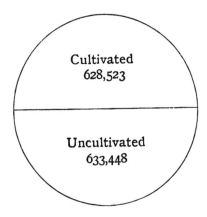

Fig. 4. Proportion of cultivated and uncultivated
areas in Aberdeenshire—practically 50 °/₀

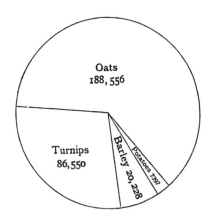

Fig. 5. Proportionate area of Crops in Aberdeenshire (1909)

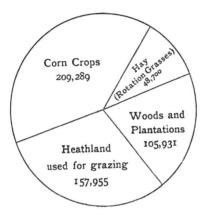

Fig. 6. Proportionate area of Crops, Pasture and Woodlands in Aberdeenshire (1909)

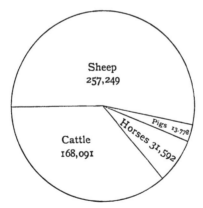

Fig. 7. Proportionate numbers of live-stock in Aberdeenshire (1909)

Fig. 8. Quantity of Fish (all kinds) landed in Aberdeenshire as compared with that of Scotland (1909), almost 50 %

Fig. 9. Quantity of Herrings landed in Aberdeenshire as compared with that of Scotland (1909)